# 绿色生态环保主题口译教程

主　编　江　雯　汪学华　李茂彪

中国水利水电出版社
www.waterpub.com.cn
·北京·

## 内 容 提 要

本书为主题型口译训练教程，旨在培养具有生态环保主题背景知识的口译人才。作者结合自身口译经验，从生态环保相关领域的真实语料中甄选和改编具有科学性和针对性的三十多篇文章作为训练材料。全书共分八个单元，内容包括生态保护、海洋与湿地、气候变化、生物多样性保护、荒野与国家公园、野生生物保护、园艺和绿色能源。

本书适用于本科英语专业、本科（硕士）翻译专业、农林院校口译课程、绿色生态主题口笔译课程或工作坊的学生。英语爱好者，尤其是关注生态环保领域口译爱好者，也可作为参考书使用。

图书在版编目（CIP）数据

绿色生态环保主题口译教程 / 江雯, 汪学华, 李茂彪主编. -- 北京 : 中国水利水电出版社, 2023.11
ISBN 978-7-5226-1928-6

Ⅰ. ①绿… Ⅱ. ①江… ②汪… ③李… Ⅲ. ①生态环境保护－英语－口译－教材 Ⅳ. ①X171.4

中国国家版本馆CIP数据核字(2023)第217423号

策划编辑：陈艳蕊　责任编辑：邓建梅　加工编辑：王一然　封面设计：苏敏

| | |
|---|---|
| 书　名 | 绿色生态环保主题口译教程<br>LÜSE SHENGTAI HUANBAO ZHUTI KOUYI JIAOCHENG |
| 作　者 | 主　编　江　雯　汪学华　李茂彪 |
| 出版发行 | 中国水利水电出版社<br>（北京市海淀区玉渊潭南路 1 号 D 座　100038）<br>网址：www.waterpub.com.cn<br>E-mail：mchannel@263.net（答疑）<br>sales@mwr.gov.cn<br>电话：（010）68545888（营销中心）、82562819（组稿） |
| 经　售 | 北京科水图书销售有限公司<br>电话：（010）68545874、63202643<br>全国各地新华书店和相关出版物销售网点 |
| 排　版 | 北京万水电子信息有限公司 |
| 印　刷 | 北京中献拓方科技发展有限公司 |
| 规　格 | 170mm×240mm　16 开本　12.25 印张　205 千字 |
| 版　次 | 2023 年 11 月第 1 版　2023 年 11 月第 1 次印刷 |
| 定　价 | 58.00 元 |

# 前　言

口译在中国作为一门职业始于两千多年前。在古代，从事口译被称为“象胥之职”。口译作为正式专门职业始于第一次世界大战“巴黎和会”。20 世纪 70 年代初，随着中国重返联合国，80 年代进行改革开放，以及加入 WTO，中国对外交流飞速发展。政治、经济、科技、商务、文旅、体育等国际交流与合作百花齐放，高质量和高层次的翻译人员供不应求，十分紧缺。基于市场需求，越来越多的国内高校开设翻译硕士专业学位（MTI），培养职业翻译人员。

十八大以来，习近平总书记提出了中国特色社会主义事业“五位一体”的总体布局，把生态文明建设放在突出地位，融入经济建设、政治建设、文化建设、社会建设各方面和全过程。在这样的时代大背景下，国内外市场对生态环境保护主题的翻译需求日渐增加。生态环保类翻译专业性较强，属于科技文体翻译。然而，具备相关领域背景知识的翻译人员较为稀缺，以生态环保为主题的翻译教材也较为匮乏。为满足翻译市场对相关领域的专业翻译人才以及对应的翻译教学训练内容的需求，编者结合自己在相关领域的翻译经验，从生态环保主题的真实语料中甄选和改编具有科学性和针对性的三十多篇文章作为训练材料，内容包括生态保护、海洋与湿地、气候变化、生物多样性保护、荒野与国家公园、野生生物保护、园艺和绿色能源。

本书在编排上希望能给学习者和教师提供灵活的选择。每个单元分为两部分，第一部分为主题口译训练，配有专业术语表。主题背景知识介绍可作为短时记忆或视译练习材料；随后是句子口译训练，用于训练短时记忆与语言重组能力；然后是篇章口译，用于实训；最后是相关概念讲解，用于丰富背景知识。所有练习内容与该单元主题相关，教师可灵活选择其中材料作为课前、课堂和课后练习任务。第二部分是口译技能讲解，涉及基本的口译听力理解、口译记忆、口译笔记三大核心技能。

在本书编写过程中，编者力求完美，但仍然难免存在疏漏和欠妥之处，恳请广大同仁和读者不吝指正，以便修正与完善。

编　者

2023 年 11 月

# 目　录

# Unit 1　Ecological Conservation
# 生 态 保 护

## Introduction

### What is Ecological Conservation?

Ecological conservation is the process of preserving the natural environment for the current and future generations. It may involve many types of preservation goals, including those aimed at protecting land, water, animals, plants, and air quality. Some ecological conservation programs are regulated by law, and require people to take certain steps to protect the planet or face legal penalties. Others are simply voluntary, and serve as recommendations for people and businesses who wish to protect the environment. One of the biggest goals of ecological conservation involves educating people about the benefits of sustainable living, as many believe that this is the best way to bring about real change.

Sustainable living is at the heart of the conservation movement. Sustainability can be defined as choosing materials and technology that provide the things people need today without impacting the survival and lifestyle of future generations. This means focusing on long-term agricultural strategies to feed people today and ensure that future generations will also have enough to eat. It also means implementing strategies that prevent people from fishing a certain species to extinction, or killing off the last of an endangered species for profit or sport. Even natural resources like fossil fuels, minerals, and metals must be protected through recycling or conservation efforts.

Part of ecological conservation also involves protecting the land, air, and water. This could mean preserving wildlife habitats or areas of exceptional beauty. Pollution controls can be implemented to keep air and water clean, and people can be taught to conserve water so that humans will always have access to safe drinking water. Ecological conservation also focuses on the ozone layer. Companies are discouraged or forbidden from using many ozone depleting substances, and investments have been made in finding more sustainable alternatives.

## Glossary

| | |
|---|---|
| greening | 绿化 |
| lighting system | 照明系统 |
| emissions cap | 废气排放上限 |
| volatile organic compounds | 挥发性有机化合物 |
| garbage classification | 垃圾分类 |
| ecological industry | 生态产业 |
| shanty town | 棚户区 |

| | |
|---|---|
| ecological degradation | 生态退化 |
| The Yangtze River Economic Belt | 长江经济带 |
| Three-North Shelter Forest Program | “三北”防护林工程 |
| The Grain for Green Project | 退耕还林工程 |
| Hong Kong - Zhuhai - Macao Bridge | 港珠澳大桥 |
| ecological corridor | 生态廊道 |
| Redlines for Ecological Conservation/ Ecological Conservation Redlines | 生态保护红线 |
| sustainable development | 可持续发展 |
| water and soil erosion | 水土流失 |
| returning farmland to forestry and grassland | 退耕还林还草 |
| crop rotation and fallow system | 轮作休耕 |
| ecological compensation mechanism | 生态补偿机制 |
| carbon neutrality | 碳中和 |
| ecological civilization | 生态文明 |

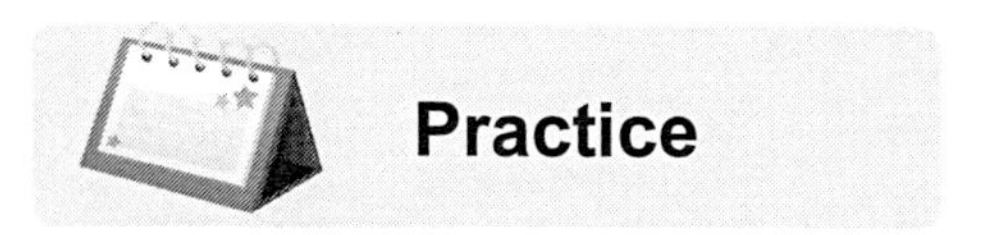

## Practice

### 1. Sentence Interpreting

（1）Some manufacturing companies are committed to developing low-carbon products and greening the overall supply chain through reduction, reuse and recycling.

（2）Energy efficiency is continually enhanced through greening the lighting system, reshaping the industrial boilers, and reusing the leftover heat from industrial processes. As a result, carbon emission has been reduced by 17.3%.

（3）The total emissions per year of each company cannot exceed an emission cap set by the government from 2008 to 2012. The emissions of major pollutants have dropped dramatically.

（4）Companies are encouraged by government to join the corporate environmental network and release an environmental action declaration to consciously undertake their corporate environmental responsibility. About 80% of the companies have joined the network and half of them have released their monthly environmental reports.

（5）While developing the economy, people in the development zone also care about their social responsibility. They have carried out a series of nonprofit activities to promote the idea of green development and eco-civilization, such as tree planting, environmental volunteering and garbage classification.

（6）为助力经济技术开发区创建生态园区，区政府将投入专项资金，以支持开发区改善和优化产业环境，确保其具备必要的条件和实力。

（7）我们将坚定不移地致力于减少挥发性有机物的排放，致力于营造“蓝天白云”的生态环境，承担起我们的环保责任，为环保事业贡献我们的力量。

（8）为了推动经济与环境的和谐共生，当地决策者决定摒弃传统的发展模式，转而以建设生态文明城市为目标。自 2007 年启动以来，该城市明确提出了建设目标：打造生态环境良好、生态产业繁荣、生态理念深入人心、文化特色鲜明、市民和谐幸福、政府廉洁高效的生态文明城市。

（9）通过大规模植树造林、清理河道垃圾、提升水处理技术和建设湿地公园等一系列措施，贵阳的城市面貌焕然一新，清洁、绿色且充满魅力，成为中国生态文明实践的典范。

（10）这片曾经作为贵阳最大棚户区的区域，经过精心规划与重建，已蜕变为一个绿意盎然的宜居城市综合体，为约 50 万居民提供了舒适的生活环境，并创造了 15 万个就业机会。

## 2. English-Chinese Interpreting

What kind of ecological civilization should China build? What challenges or obstacles lie ahead? Faced with the realities of environmental pollution, degraded ecosystems, and resource constraints, what practical policies are being enacted?

Humans and nature coexist in harmony in China, and building an ecological civilization is a fundamental strategy to achieve sustainable development. Almost every major city in China develops alongside a river. However, studies in 2013 revealed shocking statistics that China's 10 biggest waterways were polluted for over half of their combined length. Of all China's rivers, the one that raises the gravest concern is the Yangtze River. State heads have visited the river several times, calling for a major effort to protect it and a ban on large-scale infrastructure development along the river course. By prioritizing ecology and green development, the development path of the Yangtze River Economic Belt is becoming clearer. Currently, the Yangtze River is generally in good environmental condition. Also, the water quality of the Yellow River, and two other key waterways, the Heilongjiang River and Nenjiang River, has been improving.

The Three-North Shelter Forest Program has been a great success in global ecosystem governance. The Grain for Green Project has created more than 30 million hectares of forests and grasslands in the past 20 years. China's commitment to green development is contributing to the well-being of the entire world. In spring 2019, the results of a NASA environmental study, revealed that: from 2000 to 2017, China has added about a quarter of the new green area worldwide, making it the world's biggest contributor.

The main idea of sustainable development is to achieve harmony between human and nature. The builders of the Hong Kong-Zhuhai-Macao Bridge made a promise that its construction would not to force the dolphins to leave their natural habitat during its construction, and this promise has evidently been fulfilled.

Many of China's cities are being covered rapidly in green. China's environmental protection revolution is transforming the country's economic and energy structure. China has become a beautiful country, with lush mountains, clean rivers, and idyllic charm.

**3. Chinese-English Interpreting**

近年来，中国在新能源和可再生能源的利用上已跃居全球首位。据美国 NASA 的观测数据，自 2000 年以来，全球新增绿化面积的 25%归功于中国。在“美丽中国”建设的推动下，绿色发展取得了显著成效。煤炭在能源消费中的占比已降至约 56%，而清洁能源的比重则提升至 25.3%。中国的光伏、风能装机容量和发电量，以及新能源汽车的产销量均领跑全球。值得一提的是，2021 年新能源汽车销售量同比增长高达 1.6 倍，占据了世界销售量的一半以上。在碳排放方面，中国同样取得了令人瞩目的成绩。2020 年，中国单位 GDP 的二氧化碳排放相较于 2015 年下降了 18.8%，与 2005 年相比更是下降了 48.4%，这一成绩不仅超越了国际社会设定的 40%～45%的减排目标，更使中国成为全球能耗强度降低速度最快的国家之一，从而有效地扭转了二氧化碳排放快速增长的趋势。

中国的生态环境质量正在稳步提升。从 2017 年至 2021 年，全国地级及以上城市的空气质量显著改善，细颗粒物浓度下降了 25%，优良天气占比上升了 4.9 个百分点，而重污染天数则锐减近四成，蓝天白云、繁星闪烁的美景已成为人们生活中的常态。在水质方面，全国达到或优于 Ⅲ 类水体的比例已提升至 84.9%，而劣 V 类水体的比例则下降至 1.2%。在土壤安全利用方面，利用水平稳定提升，同时初步划定了生态保护红线，并持续开展大规模的国土绿化行动，森林覆盖率已达到 23.04%，为国家的生态环境增添了更多绿意。

**4. English-Chinese Interpreting**

**What you need to know about the sandstorm in Beijing?**

A sandstorm swept across Beijing and several northern regions in China on Monday. Where did the sandstorm come from? How long will it last? Let's take a look.

Q: Where did the sandstorm come from?

A: The sandstorm originated in the Gobi Desert areas of Mongolia. During the eastward movement of a Mongolian cyclone and its subsequent cold air mass, strong winds picked up sand and dust, which were then transported southward and eastward at high altitudes and eventually settled in downstream areas such as Beijing, resulting in the sandstorm.

Q: When will the dusty weather weaken? Should we expect more sandstorms this week?

A: According to the National Meteorological Center, the sandstorm will weaken starting from Tuesday. It is forecasted that another sandstorm may occur around April 14.

Q: Why are sandstorms in North China so frequent this year?

A: The cold air activity in the country has remained active in March and April. During the influence of cold air, Mongolian cyclones and cold fronts on the ground have become favorable catalysts, transporting sand and dust from the sand source areas downstream to even more southern and farther areas.

Q: Is a yellowish sky always an indicator of a sandstorm?

A: Not exactly. Only when the sandy and dusty weather reaches a certain intensity can it be called a sandstorm. In meteorology, according to the severity of the impact, the dusty weather can be divided into floating dust, blowing sand, sandstorm, strong sandstorm, and severe sandstorm.

Q: What precautions to take in dusty weather?

A: Stay indoors as much as possible, and keep windows and doors closed. Use air purifiers, sprinkle water, use humidifiers, and mop the floor with wet cloth to remove dust; Wear masks, gauze or other dust-proof items when going outdoors, and clean your face and nose when coming back home; When driving, reduce speed, turn on headlights, and drive with caution.

## Related knowledge

### Building an Ecological Civilization

Although China's remarkable economic growth has transformed its economy into an industrial powerhouse, it has also caused significant ecosystem degradation, biodiversity loss, and pollution. These crises threaten the health and safety of the current population and undermine future development. In response to these concerns, the Chinese government has developed a policy agenda of "ecological civilization" in both national planning and local management to place the environment at the core of development strategy.

Since 2007, China's central government has enacted a wide range of legislation and high-level directives under the umbrella of ecological civilization. The Circular Economy Promotion Law was promulgated in 2008 to mitigate the environmental impacts and improve resource conservation of all economic sectors. Since 2012, China has made remarkable progress in formulating and implementing plans to improve the environment and ecology of the country, along with its stellar economic progress.

For example, the Air Pollution Action Plan released in September 2013 was a major milestone in China's march to build an ecological civilization. It was a most ambitious program, the type of which was not witnessed in the history of any country. The objective was to reduce air pollution at the national level in five years. The standards for PM2.5 (particulate matter with a diameter of less than 2.5 micrometers) were set for different regions. These targets required 33% reduction for Beijing and 15% for the Pearl River Delta region by the end of the plan in 2017. When these targets were set, very few Chinese, even fewer international experts, thought they were attainable.

The targets meant Beijing had to reduce its PM2.5 level from 89.5 micrograms per cubic meter in 2013 to 60 micrograms per cubic meter in 2017. Beijing thus closed its coal-fired power stations, prohibited people in surrounding areas from burning coal for heating in winter, phased out fossil-fuel powered vehicles, ensured new cars were fuel-efficient and used other policy measures to reduce the PM2.5 level. By 2017, Beijing's PM2.5 level had been brought down to 58 micrograms per cubic meter, a 35% reduction. The Pearl and Yangtze River delta regions, too, beat their targets.

## 口译概论

### 1. 口译的定义

口译，简而言之，即口头翻译。在这一过程中，口译员将即时接收的信息以口头形式准确、迅速地由一种语言转换为另一种语言，从而在不同语言之间构建起一座沟通的桥梁，确保各方参与者能够进行高效交流。

根据口译的活动性质，它可细分为会议口译（conference interpreting）、随从口译（personal interpreting）和联络口译（liaison interpreting）等。而从工作方式来看，口译又包括同声传译（simultaneous interpreting）、交替传译（consecutive interpreting）和耳语式传译（whispered interpreting）。此外，根据口译的特定场合，还可以将其划分为法庭口译（court interpreting）、导游口译（tour guide interpreting）、电视访谈口译（interview interpreting）和商务口译（business interpreting）等多种类型。

### 2. 口译与笔译的区别

（1）工作环境的不同。口译工作具有鲜明的现场性和严格的时限性。现场性意味着口译员必须在公众面前进行即时翻译，且无法自主选择工作地点；而笔译的工作地点则无限制。时限性则指口译员在翻译过程中一般不能对已经翻译的内容进行大量更正或补充，除非发现重大错误或遗漏。换言之，口译员通常只有一次机会进行翻译，频繁的修改会影响交流和沟通的流畅性、效率，甚至可能影响客户对口译服务的满意度。相比之下，笔译工作则允许译者拥有充足的时间去反复阅读原文、深入理解其内涵，并在完成初稿后，与委托方就相关专业内容进行沟通，进行反复的修改和润色，直至双方都满意为止。

（2）工作方法不同。口译员在工作时，特别是进行会议交替传译和陪同口译时，通常需独自应对，且无法即时查阅词典或其他参考资料。而笔译员则享有更多的自由，他们可以随时查阅各种工具书和专业文献，反复推敲，字斟句酌，甚至咨询相关领域的专家以获取帮助。

（3）反馈形式不同。口译是一种即时的翻译活动，口译员直接面对发言人和听众。他们能够即时从对方的面部表情、情绪、手势等身体语言中获取反馈，并根据现场情况和听众的反应灵活调整自己的音量、语速和用词。而笔译员在翻译过程中则较少有机会与原文作者或读者进行面对面的交流，更无法实时根据他们的反馈进行修正。

（4）质量标准各有侧重。笔译遵循的是“信、达、雅”的原则，即译文要忠实于原文，表达流畅，并具有一定的文学性。而口译则更强调“准、顺、快”的标准。这里的“准”即准确，指的是准确理解并即时传达源语信息；“顺”即通顺，要求口译员在表达时语言流畅，符合目的语的表达习惯；“快”即快速反应，意味着口译员需要具备迅速反应的能力，例如，在交替传译中，发言人发言结束后，口译员应立即开始翻译。

### 3. 口译过程与训练内容

（1）口译过程。口译可大致划分为三个连贯的步骤：理解—记忆—表达。

首先是理解，即准确捕捉发言人的讲话核心意义和深层意图。这一步骤要求口译员具备扎实的语言能力、专业知识和背景知识，以确保能从零散的词汇和句子中提炼出连贯且逻辑清晰的意义。

其次是记忆，即短时记忆，涉及将理解的信息暂时储存在脑海中或记录于笔记中。记忆以理解为基石，缺乏理解，记忆的内容将变得琐碎而无意义。

最后是表达，即以目标语言自然流畅地重现话语的深层含义。

（2）训练内容。基于“理解—记忆—表达”的核心步骤，口译技巧的训练可细化为三个关键领域：理解技巧训练、记忆技巧训练和表达技巧训练。

第一，理解技巧训练，包括主动听取信息（active listening）和口音辨别。

主动听取信息：口译员需通过听觉和视觉双重渠道接收信息，其中听觉是核心。训练时，应着重于捕捉关键信息、识别逻辑关系以及提炼主旨，确保注意力始终高度集中。

口音辨别：除了具备卓越的听力和理解能力外，口译员需熟悉并辨识不同国家和地区的口音。随着“一带一路”倡议的推进，特别是在云南等边境省份，与南亚和东南亚国家的交流日益增多。目前，在云南的各种国际会议中，具有纯正的英语口音母语背景的发言人越来越少。相反，来自印度、孟加拉国、尼泊尔、越南、老挝、柬埔寨等国的发言人越来越多。很多发言人的英语口语受母语的表达方式和发音影响很大，因此，熟悉以上国家或地区的口音，掌握该地区的英语口音特点是很有必要的。

第二，记忆技巧训练，包括短期记忆训练和笔记法训练。

短期记忆训练：记忆分为短期记忆和长期记忆。口译员需具备良好的短期记忆能力，即在口译过程中，没有笔记的情况下能够记住发言人的讲话主旨内容。通过专项短期记忆训练，增强口译员在没有笔记辅助的情况下，短时间内准确记忆并复述信息的能力。

笔记法训练：笔记法是口译中不可或缺的信息记录方式，通过训练，口译员能够迅速、准确地记录难以凭短期记忆保留的关键信息，如数字、专有名词和逻辑框架等。

第三，表达技巧训练。训练目标应达到“准、顺、快”的标准，即信息传达准确、语言表达流畅、反应迅速。为此，需要进行演说技巧、语言重组技能、跨文化交际分析和应对策略等多方面的综合的表达技巧训练，以确保信息能够自然、高效、灵活、简洁、准确、完整地传达给听众，并且能突出信息的重点和意图。

**1. Sentence Interpreting**

（1）部分制造业企业致力于开发低碳产品，并通过减少（废弃物）、再利用和回收策略来实现整个供应链的绿色化。

（2）通过照明系统的绿色化改造、工业锅炉的技术更新以及工业过程余热的循环利用，不断提高了能源效率，减少了17.3%的碳排放量。

（3）2008—2012年，各企业严格遵守国家规定的排放上限，使主要污染物排放量大幅下降。

（4）政府鼓励企业加入企业环境网络，共同发布环境行动宣言，自觉承担企业环保责任。约80%的企业加入该网络，其中半数企业定期发布环境月报。

（5）在发展经济的同时，开发区里的人员也承担自己的社会责任。他们开展了一系列公益活动，如植树造林、环保志愿服务和垃圾分类等，旨在推广绿色发展和生态文明的理念。

（6）In order to support the economic and technological development zone to create an ecological park district, the district government will appropriate part of the financial resources for the development zone, so that it will have the conditions and ability to optimize the industrial environment conditions.

（7）We will, as always, persist in reducing the emission of volatile organic compounds, strive to keep the sky blue with white clouds, and shoulder our responsibilities and contribute our strength.

（8）In order to promote the harmony between economic development and environmental protection, local policy-makers decided to abandon traditional development model and to build an ecologically civilized city. The city's ecological civilization construction was officially launched in 2007, and a description of the ecological civilization city was made. It's goal is to build an ecologically civilized city with a good ecological environment, developed eco-industries, people's strong ecological awareness, distinctive cultural characteristics, harmonious and happy citizens, and a clean and efficient government.

（9）By supporting large-scale afforestation campaigns, removing waste from rivers, improving water treatment technology, and building wetland parks, Guiyang has become cleaner, greener, and more attractive, and has become a beacon of ecological civilization practices in China.

（10）It used to be the largest shanty town of Guiyang, but after the reconstruction project, it has been transformed into a green and comfortable urban complex, where about 500,000 residents live, and 150,000 jobs are created.

## 2. English-Chinese Interpreting

中国致力于建设怎样的生态文明？前方有哪些挑战或障碍？面对环境污染和生态系统退化以及资源约束的现实，正在制定哪些政策？

人与自然和谐相处，建设生态文明是实现可持续发展的根本方略。中国的各大城市多依河而建。然而，2013 年研究揭示了一个令人震惊的统计数据，即中国十大水道的总长度的一半以上遭受了污染。在中国的所有河流中，最受关注的是长江。国家领导人曾多次视察长江，并强调要大力保护，禁止沿途进行大规模建设。如今，坚持生态优先、绿色发展，长江经济带发展路径日趋清晰，长江总体状况良好。同时，黄河、黑龙江、嫩江重点水道水质不断改善。

在全球生态系统治理方面，中国的三北防护林工程取得了巨大成功。20 年来，退耕还林工程创造了超过 3,000 万公顷的森林和草原。中国坚持绿色发展，正在造福世界。2019 年春，美国宇航局的一项环境研究指出：2000—2017 年，中国新增绿地面积占全球新增绿地面积的四分之一左右，成为全球最大的贡献者。

可持续发展的核心是人与自然和谐共生。港珠澳大桥的建设就是一个生动的例子。建设者承诺，大桥的建设不会迫使海豚离开它们的自然栖息地，这一承诺已经得到了兑现。

如今，中国的许多城市的绿化率迅速提升。环保革命正在改变国家的经济和能源结构。我们有理由相信中国将是一个山清水秀、如田园般美丽的地方。

**3. Chinese-English Interpreting**

In recent years, China has become the world's largest country in the use of new and renewable energy. According to the observation data of NASA, about 25% of the newly added green area in the world since 2000 came from China. Over the past few years, the building of Beautiful China has made staged progress. The effect of green development has gradually emerged. The proportion of coal consumption in energy consumption has dropped to about 56%, and the proportion of clean energy has risen to 25.3%. China ranks the first in the globe in installed capacity of solar photovoltaic (PV) and wind power, electricity generated from PV and wind power, and sales of new energy vehicles. Production and sales of new energy vehicles in 2021 increased by 1.6 times year-on-year, accounting for more than half of the global sales. Also, China has made remarkable achievements in carbon emissions reduction. In 2020, China's carbon dioxide emissions per unit of GDP decreased by 18.8% compared with 2015, and by 48.4% compared with 2005, exceeding the 40%-45% target set by the international community. It is also one of the countries with the fastest reduction in energy consumption intensity in the world, reversing the rapid growth of carbon dioxide emissions.

The quality of China's ecological environment continues to improve. From 2017 to 2021, Air quality improved significantly, the concentration of fine particulate matter in cities at prefecture level and above across the country dropped by 25%, the proportion of days with good air quality increased by 4.9%, and the number of days with heavy pollution dropped by nearly 40%. Blue sky, white clouds and twinkling stars have returned. In terms of water quality, the proportion of water bodies meeting or higher than Class III rose to 84.9% nationwide, and the proportion of water bodies inferior to Class V dropped to 1.2%. In light of soil safety and use, the level of safe use of soil has been steadily improved, the red line for ecological conservation has been initially delineated, and large-scale greening actions have been carried out continuously, the forest cover has reached 23.04%, providing more greenery to the country's ecological environment.

## 4. English-Chinese Interpreting

### 北京沙尘暴的问与答

本周一，一场沙尘暴席卷了北京和中国北方的几个地区。沙尘暴从何而来?它会持续多久? 让我们来了解其中缘由。

问：沙尘暴从何而来?

答：沙尘暴起源于蒙古的戈壁沙漠地区。在蒙古气旋及其后的冷气团东移过程中，强风卷起沙尘，在高空向南、向东移动，最终落在北京等下游地区，形成沙尘暴。

问：沙尘天气什么时候会减弱？本周会有更多的沙尘暴吗？

答：据国家气象中心消息，从周二开始，沙尘暴将减弱。据预测，4 月 14 日前后可能会出现另一场沙尘暴。

问：今年华北沙尘暴为何如此频繁？

答：三四月份全国冷空气活动依然活跃。在冷空气影响期间，地面上的蒙古气旋和冷锋成为有利的催化剂，将沙尘从沙源区向下游输送到更南、更远的地区。

问：淡黄色的天空一定是沙尘暴的征兆吗？

答：不完全是。只有当沙尘天气达到一定强度时，才能称为沙尘暴。在气象学上，根据影响的严重程度，沙尘天气可分为浮尘、扬沙、沙尘暴、强沙尘暴和特强沙尘暴。

问：沙尘天气有哪些防护措施？

答：尽量待在室内，关好门窗；（为了改善室内空气质量）可以使用空气净化器或洒水，使用加湿器以及用湿布拖地以减少空气中的灰尘；外出佩戴口罩、纱布等防尘物品，回家后及时清洁面部，尤其是鼻子；开车时，要降低车速，打开前照灯，谨慎驾驶。

# Unit 2 Oceans and Wetlands
# 海洋与湿地

### What is a Wetland?

Wetlands are areas where water covers the soil, or is present either at or near the surface of the soil all year round or for varying periods of time during the year, including during the growing seasons. Water saturation (hydrology) largely determines how the soil develops and the types of plant and animal communities living in and on the soil. Wetlands may support both aquatic and terrestrial species. The prolonged presence of water creates conditions that favor the growth of specially adapted plants (hydrophytes) and promote the development of characteristic wetland (hydric) soils.

Wetlands vary widely because of regional and local differences in soils, topography, climate, hydrology, water chemistry, vegetation and other factors, including human disturbances. Indeed, wetlands are found from the tundra to the tropics and on every continent except Antarctica. Two general categories of wetlands are recognized: coastal or tidal wetlands and inland or non-tidal wetlands.

Many of these wetlands are seasonal (they are dry in one or more seasons every year), and, particularly in the arid and semiarid west, may be wet only periodically. The quantity of water present and the timing of its presence in part determine the functions of a wetland and its role in the environment. Even wetlands that appear dry at times for significant parts of the year–such as vernal pools–often provide critical habitat for wildlife adapted to breeding exclusively in these areas.

## Glossary

| | |
|---|---|
| nuclear power station/plant | 核电站 |
| radioisotope | 放射性同位素 |
| concentration | 浓度 |
| tritium | 氚 |
| atom isotope | 原子同位素 |
| International Atomic Energy Agency (IAEA) | 国际原子能机构 |
| vaporize | 蒸发 |
| food chain | 食物链 |
| tundra | 冻原、苔原 |
| gene pool/bank | 基因库 |
| crustacean | 甲壳纲动物 |
| amphibian | 两栖动物 |
| mammals | 兽类/哺乳类 |
| underground water | 地下水 |
| domestic sewage | 生活污水 |

| | |
|---|---|
| industrial sewage | 工业废水 |
| agricultural sewage | 农业污水 |
| carbon sink | 碳汇 |
| ocean acidification | 海洋酸化 |
| plastic pollution | 塑料污染 |
| hydropower energy | 水能电源 |
| per capita | 人均 |
| reclamation | 开垦 |
| wetland nature reserve | 湿地自然保护区 |
| Ramsar Sites | 拉姆萨尔湿地 |
| climate indicator | 气候指标 |
| World Meteorological Organization | 世界气象组织 |
| ocean trench | 海沟 |
| dead zone | 死区、死亡区 |
| Sustainable Development Goals (SDGs) | 可持续发展目标 |
| the Paris Agreement | 《巴黎协定》 |

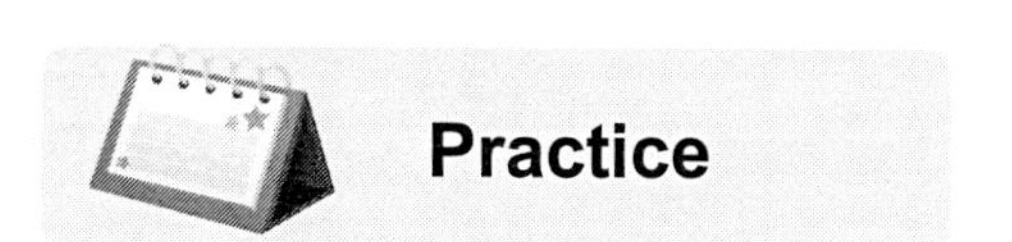

## Practice

### 1. Sentence Interpreting

（1）As of early April, the amount of nuclear contaminated water stored in the tanks at the Fukushima nuclear power station in Japan was 1.25 million tons, with the amount of radioactive water still increasing by about 140 tons a day.

（2）It is predicted that as of March 2023, the volume of the contaminated water will hit 1.37 million tons, reaching the upper limit of the current water tanks Japan has built for the project.

（3） However, that the Japanese government has decided to discharge the nuclear-contaminated water into the sea after diluting it to make the radiation level lower than its national standards deserves the vigilance of the whole world.

（4）Nuclear biochemical protection experts from different countries have said that it is almost impossible to eliminate all the radioisotopes from the contaminated water with the current technology, and to make the concentration of the tritium lower than the national standard of Japan does not mean the radioactive elements will have been removed.

（5）The remaining radioisotopes in the water will accumulate in the food chain of organisms, which is a serious problem. How long the pollution will affect the marine environment remains unknown, and once the nuclear-contaminated water is discharged into the sea, the speed of its spread, its effects and environmental risks will all be in an uncontrollable state.

（6）日本在处理和报告水体净化过程及其结果时，应当坚守公开、透明和科学的原则。同时，日本应主动接受国际组织，特别是国际原子能机构以及相关国家的第三方监督。这种监督将确保全球能够及时了解水体的放射性状况。

（7）日本是受到核污染水影响的一方。经过多年的艰辛努力，福岛周边的渔业秩序才恢复正常运营。因此，福岛周边从事捕鱼作业的渔民以及相关团体，作为最直接受影响的人群，他们坚决反对将核污染水排入海洋的计划。

（8）污水处理无疑已成为全球共同关注的重大问题。在处理污水时，我们面临两种选择：一是将其排入大海，而另一种则是采用蒸气释放技术。第一种方案由日本专家小组推荐，从经济和技术的角度来看，它被普遍认为是更为可行和高效的解决方案。

（9）在核电安全问题上，人类应该是一个命运共同体，利益与风险共存。

（10）海洋哺乳动物广泛分布于全球的海洋生态系统中，它们构成了一个多样化的群体，身体拥有独特的适应性。

**2. English-Chinese Interpreting**

**Message by UN Secretary-General António Guterres on World Oceans Day 8 June 2022**

Last month, the World Meteorological Organization revealed that four key climate indicators broke new records in 2021: sea level rise, ocean heat, ocean acidification, and greenhouse gas concentrations. It is clear that the triple crises of climate change, biodiversity loss and pollution are threatening the health of our oceans, on which we all ultimately depend.

The ocean produces more than 50% of the planet's oxygen and is the main source of sustenance for more than a billion people. Ocean-based industries employ some 40 million people. Yet ocean resources and biodiversity are being undermined by human activities. More than one third of the world's fish stocks are harvested at biologically unsustainable levels. A significant proportion of coral reefs have been destroyed. Plastic pollution has reached the remotest islands and deepest ocean trenches. Coastal death zones from land-based pollution are growing.

It is time to realize that, to achieve the Sustainable Development Goals (SDGs) and the objectives of the Paris Agreement on Climate Change, we urgently need collective action to revitalize the ocean. That means finding a new balance in our relationship with the marine environment. It means working together with nature, not against it, and building inclusive and diverse partnerships across regions, sectors and communities to collaborate creatively on ocean solutions.

Ensuring a healthy and productive ocean is our collective responsibility, which we can only fulfill by working together. On this World Oceans Day, I urge all those with a stake in ocean health to come together to revitalize our seas and oceans.

### 3. Chinese-English Interpreting

#### 为什么湿地被称为“地球之肾”？

湿地，顾名思义，即水分充沛的湿润之地。全球湿地的总面积超过 800 万平方千米，约占地球陆地面积的 6%。作为陆地与水域的交汇地带，湿地不仅孕育了丰富的陆生和水生动植物资源，更构筑了一个独一无二的天然基因库，这是任何其他单一生态系统所无法比拟的。其特有的水文、土壤和气候条件为复杂而完善的动植物群落提供了理想的生长环境。

湿地，作为地球上生物多样性最为丰富的区域之一，为约 20%的生物物种提供了独特的生存环境。这里不仅是甲壳类、鱼类、两栖类、爬行类、兽类等多种生物的繁衍之地，也是珍稀鸟类栖息和繁衍的理想场所。湿地与森林、海洋共同构成了地球上的三大生态系统。湿地的一个重要功能是调节水分平衡。湿地被誉为“天然海绵”，在洪水泛滥时，它能吸收并储存大量水分，避免水患；而在干旱时期，湿地又能释放储存的水分，为周边河流和地下水提供水分补给。

湿地与河流显著不同，其水流速度缓慢。因此，当生活污水、工业废水和农业污水流入湿地时，由于流速减缓，其中的有毒和有害物质能够逐渐沉淀。此外，湿地植物具备强大的吸收和分解能力，它们能够吸收有毒物质，并通过生物转化过程将其分解，从而有效净化水质。湿地对地球的作用与人体中的肾脏相似，都起到了调节水分循环和排泄废物的重要作用。因此，湿地被誉为“地球之肾”。然而，随着城市化的快速推进和农业开垦的加剧，全球湿地的面积正以惊人的速度减少。保护湿地，维护其生态平衡，已经成为我们刻不容缓的责任。

### 4. English-Chinese Interpreting

Unfortunately, different types of wetlands, such as lakes, rivers, marshes, and coasts, are among the most heavily degraded of all ecological systems worldwide. Half of them may have been destroyed in the past 100 years alone. Wetlands are not only essential for ecological functions such as maintaining biodiversity, controlling floods, and the removal of pollutants, but also serve necessary economic functions in rice and fish production, transport, and hydropower energy.

With one-fifth of the world's population, conservation of wetlands in China is crucial. Although China ranks fourth in the world in wetland surface, it ranks 109th in the world in fresh water per capita. It is host to some of East Asia's most important wetlands such as Poyang Lake and Asia's longest river, the Yangtze River. These wetlands in China are not only the birthplace of major rivers such as the Lancang-Mekong River, but also an important part of the ecosystem of population growth and economic development in the past 40 years, some 13% of China's lakes have disappeared, half of China's coastal wetlands have lost to reclamation and 50% of China's cities have insufficient drinking water that meet required hygiene standards.

Since the 1970s, when China's first wetland nature reserve was created, the number of wetland nature reserves has grown to 262, with an area of about

16 million hectares protected. China acceded to the Convention on Wetlands of International Importance on July 31, 1992 and designated seven wetland sites as the Wetlands of International Importance, known as Ramsar Sites. In October 2000, the State Forestry Administration (SFA), the governmental ministry were appointed to implement the Convention, together with 17 other government ministries and academic institutions, formulated China's National Wetland Conservation Action Plan (2000). The Action Plan serves as a guideline for the conservation and wise use of wetland resources in the country.

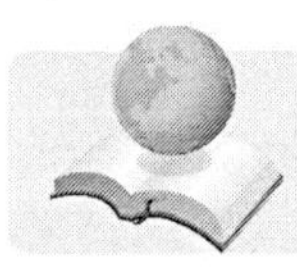

## Related knowledge

### The Ramsar Convention (The Convention on Wetlands of International Importance)

The Convention on Wetlands is the intergovernmental treaty that provides the framework for the conservation and wise use of wetlands and their resources.

The Convention was adopted in the Iranian city of Ramsar in 1971 and came into force in 1975. Since then, almost 90% of UN member states, from all the world's geographic regions, have acceded to become "Contracting Parties".

The Convention's mission is "the conservation and wise use of all wetlands through local and national actions and international cooperation, as a contribution towards achieving sustainable development throughout the world".

Wetlands are among the most diverse and productive ecosystems. They provide essential services and supply all our fresh water. However, they continue to be degraded and converted to other uses.

The Convention uses a broad definition of wetlands. It includes all lakes and rivers, underground aquifers, swamps and marshes, wet grasslands, peatlands, oases, estuaries, deltas and tidal flats, mangroves and other coastal areas, coral reefs, and all human-made sites such as fish ponds, rice paddies, reservoirs and salt pans.

Under the “three pillars” of the Convention, the Contracting Parties commit to: work towards the wise use of all their wetlands; designate suitable wetlands on the list of Wetlands of International Importance (the Ramsar List) and ensure their effective management; cooperate internationally on transboundary wetlands, shared wetland systems and shared species.

## 口译理解与短时记忆训练

记忆是对信息的接收、存储和提取。它通常分为长时记忆和短时记忆两种形式。

长时记忆存储着我们生活中长期积累的各种信息，如基础知识、个人经验、先前学习的背景知识以及专业领域的术语等。这些信息在我们需要时可以被提取出来，用于支持我们的决策、思考和学习。

短时记忆指短时间内储存于大脑中的信息，例如，一两分钟或几分钟内的语言交流内容。在口译中，短时记忆训练尤为重要，因为口译员需要在短时间内准确理解和记忆源语言的内容，并随后将其转化为目标语言进行表达。

在口译领域，虽然短时记忆训练是关键，但长时记忆，即背景知识的积累同样重要。一个拥有丰富背景知识的口译员能够更快速、更准确地理解和表达复杂的信息内容。因此，在口译训练中，我们应该同时注重长时记忆和短时记忆的训练，以提高整体的口译能力。

值得注意的是，短时记忆的实质并非简单地记忆词汇和句子，而是对信息和意义的记忆。这意味着在口译中，我们不仅要关注语言的外在形式，更要深入理解其背后的含义和逻辑关系。现在，我们来看一组关于长时记忆信息的案例。

长时记忆信息案例：

The Sun is situated at the center of the solar system.

汉译：太阳位于太阳系的中心。

The Nile River is the longest river in the world.

汉译：尼罗河是世界上最长的河流。

Mali is an African country and Ecuador is an American country.

汉译：马里是个非洲国家，而厄瓜多尔是个美洲国家。

以上四个案例均属于长时记忆信息，涉及自然和地理常识，在口译输入和输出过程中能减少记忆负担，有效帮助口译员理解全文信息。因此，口译员除了训练短时记忆，还需要不断学习和积累百科知识（to know something about everything）。以下是短时记忆信息案例。

短时记忆信息案例：

The shop assistant was dismissed as she was accused of cheating customers. (12 words)

汉译：商店的售货员被解雇了，因为她欺骗顾客而被起诉。

World Health Organization says diseases linked to smoking kill at least 2,500,000 people each year. (15 words)

汉译：世界卫生组织称每年至少有 250 万人死于与吸烟相关的疾病。

Obama had a 3-day visit to China from 15th to 18th November, 2009 and he visited the Forbidden City and the Great Wall. (24 words)

汉译：2009 年 11 月 15 日到 18 日，奥巴马在中国进行了三天的访问，参观了故宫和长城。

以上三个案例均属于短时记忆信息，里面涉及专有名词、数字和日期，凭借背景知识和（或）常识是无法记住的，需要调动短时记忆能力记住信息，才能复述。

### 1. 短时记忆的重要性

口译具有现场性和时限性的特点，也就意味着：

（1）源语（source language）信息的发布具有一次性与瞬时性特征，这就要求译员在短时间内完成对信息的加工处理。

（2）目标语（target language）的产出需要在短时间内进行，这也要求译员必须具有短时记忆的技能。

因此，短时记忆对于口译的质量至关重要。

**2. 短时记忆要记什么？**

（1）main ideas 主要观点。

（2）supporting details 支撑细节。

（3）logical links 逻辑联系。

例如：Why do people enjoy challenges? There are probably many reasons. One of them is the personal feeling of success. Someone asked Edmund Hillary, one of the first people who climbed Mount Qomolangma, "Why do you risk your life just to stand on a point of land that is a little bit higher?" His answer was climbing gave him a feeling of victory and conquering over the mountain and over his weakness.

分析：

Main idea: The reason why people enjoy challenges–feeling of success.

Supporting detail: Edmund Hillary.

Logical link: cause and effect.

**3. 难点**

（1）专有名词。

例如：President <u>Obama</u> and his <u>Cuban</u> counterpart <u>Raul Castro</u> met face to face on the sidelines of <u>the Summit of the Americas in Panama</u>. (23 words)

汉译：奥巴马总统和古巴领袖劳尔·卡斯特罗在巴拿马美洲峰会的非正式会议期间进行了会面。

分析：该句中出现的专用名词有：President Obama、Raul Castro、the Summit of the Americas、Panama，其中人名、地名、会议名称等都会阻碍我们的短时记忆。

（2）信息的顺序。

例如：The economic strength of this country is surpassed by America, China, Japan, France, and Germany, and its standard of living is falling back in the league table. (27 words)

汉译：这个国家的经济实力已落后于美国、中国、日本、法国以及德国，它的生活水平在位次表中也在下降。

分析：该句中出现了五个按经济总量位次排序的国家，这会增加我们短时记忆的负担和难度。

（3）数字。

例如：China has 271 million online consumers, meaning that almost half of China's 591 million Internet users buy products online. (19 words)

汉译：中国有 2 亿 7,100 万网络消费者，这意味着在中国的 5 亿 9,100 万网民中有一半多在网上购物。

分析：这句话虽不长，但是出现了两个数字，这也会增加短时记忆的难度。

（4）信息的密度。

例如：Problems such as the population explosion, ecological imbalance, the shortage of natural resources and environmental pollution have become major factors keeping human society from being further developed. (27 words)

汉译：人口爆炸、生态失衡、自然资源短缺、环境污染等问题都已经成为阻碍人类社会向前发展的主要因素。

分析：这四个短语 population explosion, ecological imbalance, the shortage of natural resources and environmental pollution 放在一起增加了信息的密度，在口译时会给译员的短时记忆增加负担。

（5）缺乏相关的背景知识。

例如：One dollar for changing a new part sounds reasonable but the $29,999 is exactly the value of wisdom and knowledge. (21 words)

汉译：花一美元换一个新零件听起来很合理，但是 29,999 美元才是智慧和知识的价值。

分析：这句话需要背景信息，才会理解的透彻。缺乏背景信息会导致口译短时记忆负担增加。

背景：Once there was something wrong with a machine in a factory. The workers could do nothing but to turn to a German engineer for help, who drew a line on it and said, "Take off the part where the line is drawn and replace it with a new one." The machine, with the new part on, began operating again. The engineer then asked for $30,000 for what he had done. The following were what he wrote on a bill, "The new part costs only one dollar, and knowing where the problem is with the machine costs $29,999."

译文：

一家工厂的机器出了问题。工人们无能为力，只能求助于德国工程师，后者在故障机上面画了一条线，并说："取下生产线条所在的部件，换一个新的。"新换零件的机器开始正常运行。然后，工程师索要三万美元作为他的工钱。以下是他在账单上写的内容："新零件仅花费 1 美元，但知道机器的问题出在哪里值 29,999 美元。"

## 1. Sentence Interpreting

（1）截至 4 月初，日本福岛核电站的储罐中存放着 125 万吨核污染水，而且每天还在增加约 140 吨。

（2）预计到 2023 年 3 月，福岛核电站的核污染水储存量将达到 137 万吨，这将触及目前为该项目建造的水箱的上限。

（3）然而，日本政府决定将海水稀释后排放入海，使辐射水平低于其“国家标准”，这值得全世界警惕。

（4）各国核生化防护专家指出，当前技术难以彻底消除受污染水中的放射性同位素，即使通过处理使氚浓度降至日本“国家标准”以下，也不代表放射性元素已被完全去除。

（5）水中残留的放射性同位素在生物食物链中的累积效应是一个严重的问题。这种污染对海洋环境的长远影响尚难以预测，因为一旦核污染水排入大海，其传播速度、影响范围以及潜在的环境风险都将变得不可控。

（6）Japan should report the process and results of water treatment in an open, transparent and scientific manner, and accept third-party supervision from international organizations, such as the International Atomic Energy Agency and relevant countries, so that the world will know the radioactive level of the water.

（7）Japan is also a victim of contaminated nuclear water. After so many years of hard work, the fishing industry around Fukushima is only just beginning to return to normal, so the fishermen and groups fishing around Fukushima are the firmest opponents.

（8）Sewage treatment is indeed a worldwide problem. Discharging into the sea is one of two options, the other being vapor release. The first option was recommended by a panel of Japanese experts and was deemed more economically and technically feasible.

（9）On the issue of nuclear power safety, mankind should be a community with a shared future, where benefits and risks coexist.

（10）Marine mammals are found in marine ecosystems around the globe. They are a diverse group with unique physical adaptability.

**2. English-Chinese Interpreting**

**联合国秘书长安东尼奥·古特雷斯世界海洋日致辞**

**2022年6月8日**

上个月，世界气象组织指出，2021 年海平面上升、海洋热量、海洋酸化和温室气体浓度四项关键气候指标均刷新了历史纪录。这三大危机——气候变化、生物多样性丧失和污染——正日益加剧，对人类赖以生存的海洋的健康构成严重威胁。

海洋不仅是地球上超过 50%氧气的生产者，还是 10 多亿人的主要生存来源，同时支撑着大约 4000 万人的海洋产业就业。然而，海洋资源和生物多样性正遭受人类活动的严重破坏。全球三分之一以上的鱼类种群正遭受着生物学上不可持续的捕捞压力，珊瑚礁也遭受了大规模的破坏。此外，塑料污染已经扩散至地球上最偏远的岛屿和最深的海沟，而陆源污染则导致沿海死水区不断扩大。

现如今，我们必须深刻认识到，为了达成可持续发展目标以及《巴黎协定》中关于气候变化的宏伟目标，我们迫切需要采取集体行动，以振兴并修复海洋。

这要求我们重建与海洋环境的关系，寻求一种全新的平衡。这种平衡意味着我们要与大自然和谐共生，而非对抗。同时，我们还需在各个区域、部门以及社区之间，建立起包容且多元化的伙伴关系，以创新和合作的方式，共同解决海洋问题。

确保海洋的健康与生产力是我们共同肩负的责任，而合作是实现这一目标的唯一途径。在世界海洋日这个特殊的日子里，我强烈呼吁所有关心海洋健康的利益攸关者携手并进，共同努力，修复我们的海洋。

## 3. Chinese-English Interpreting

### Why are Wetlands Called the Kidneys of the Earth?

Wetlands, as the name suggests, are places rich in water and moisture. The total area of wetlands in the world is about 8 million square kilometers, accounting for about 6% of the earth's land area. Wetlands are transitional areas between land and water, and are rich in both terrestrial and aquatic animal and plant resources, forming a natural gene pool unrivalled by any other single ecosystem. Its unique hydrology, soils and climate provide an ideal growing environment for a complex and well-established flora and fauna.

Wetlands, as one of the areas with richest biodiversity on earth, provide the living environment for 20% of the earth's biological species. All kinds of creatures, such as crustaceans, fish, amphibians, reptiles, and mammals, can reproduce here, and of course wetlands are also suitable habitats of rare birds. They, along with forests and oceans, are known as the three major ecosystems of the earth. An important role of wetlands is to regulate water balance. Some people compare wetlands to natural sponges. When floods come, wetlands can hold a lot of water. The surface of the wetlands is submerged, and the underlying soil can fully absorbs water. In times of drought, the water captured in the wetland flows out and becomes a water source to replenish the rivers and groundwater.

Wetlands are different from rivers. The water flow here is slow. After domestic sewage, industrial wastewater and agricultural sewage are discharged into the wetland, the speed of movement slows down, and the toxic and harmful substances in it will gradually deposit. Wetland plants also have the ability to absorb toxic substances. They can disintegrate toxins, and can effectively purify water. The kidneys in the human body play the role of regulating the body's water circulation and excreting metabolic waste, which is similar to the services of wetlands to the earth, so wetlands are called the kidneys of the earth. With urban development and land reclamation for agricultural use, the world's wetlands are disappearing at an alarming rate, and the conservation of wetlands is urgent.

**4. Chinese-English Interpreting**

不幸的是，湿地——包括湖泊、河流、沼泽和海岸——已成为全球生态系统中退化最为严重的类型。全球一半湿地可能在过去的仅仅 100 年内就遭到了破坏。湿地不仅具有维持生物多样性、调控洪水、清除污染物等关键生态功能，还在水稻种植、渔业生产、交通运输以及水电能源等领域发挥着不可或缺的经济作用。

中国拥有世界五分之一的人口，保护湿地资源显得尤为关键。尽管中国湿地面积位居世界第四，但人均淡水资源却仅排在第 109 位。中国还拥有一些东亚地区最重要的湿地，如鄱阳湖和亚洲最长的河流——长江，这些湿地不仅是澜沧江—湄公河等主要河流的发源地，更是生态系统的重要组成部分。然而，近 40

年来，由于人口增长和经济发展的巨大压力，中国约 13% 的湖泊已经消失，近半数的滨海湿地被填海造田用于其他用途。此外，50%的城市面临着卫生达标的饮用水短缺的问题。

自上世纪 70 年代中国首个湿地自然保护区成立以来，此类保护区已迅速扩展至 262 个，覆盖面积高达 1,600 万公顷。1992 年 7 月 31 日，中国正式加入《国际湿地公约》，其中 7 处湿地更是荣登《国际重要湿地名录》，即《拉姆萨尔湿地公约》。为切实履行公约义务，2000 年 10 月，国家林业局联合其他 17 个政府部委和学术机构共同制定了《中国湿地保护行动计划（2000）》。这份纲领性文件将为中国未来湿地的保护和合理利用提供明确的指导方向。

# Unit 3 Climate Change
# 气 候 变 化

### What is Climate Change?

Climate change refers to long-term shifts in temperatures and weather patterns. These shifts may be natural, such as through variations in the solar cycle. But since the 1800s, human activities have been the main driver of climate change, primarily due to burning fossil fuels like coal, oil and gas.

Burning fossil fuels generates greenhouse gas emissions that act like a blanket wrapped around the Earth, trapping the sun's heat and raising temperatures.

Examples of greenhouse gas emissions that are causing climate change include carbon dioxide and methane. These come from using gasoline for driving a car or coal for heating a building, for example. Clearing land and forests can also release carbon dioxide. Landfills for garbage are a major source of methane emissions. Energy, industry, transport, buildings, agriculture and land use are among the main emitters. And emissions continue to rise. As a result, the Earth is now about 1.1℃ warmer than it was in the late 1800s. The last decade (2011—2020) was the warmest on record.

Many people think climate change mainly means warmer temperatures.

But temperature rise is only the beginning of the climate change. Because the Earth is a system, where everything is connected, changes in one area can influence changes in all others.

The consequences of climate change now include, among others, intense droughts, water scarcity, severe fires, rising sea levels, flooding, melting polar ice, catastrophic storms and declining biodiversity.

## Glossary

| | |
|---|---|
| greenhouse effect | 温室效应 |
| green washing | 漂绿 |
| overpopulation | 人口过剩 |
| carbon dioxide | 二氧化碳 |
| methane | 甲烷 |
| nitrous oxide | 一氧化二氮 |
| fossil fuel | 化石燃料 |
| renewable energy | 可再生能源 |
| resilience | 韧性、适应能力、恢复力 |
| carbon footprint | 碳足迹 |
| mangrove | 红树林 |
| The UN Climate Change Conference | 联合国气候变化大会 |
| landfill | 垃圾填埋 |
| family | 属 |
| Just Energy Transition Partnerships | 公正能源转型伙伴关系 |

| | |
|---|---|
| climate finance | 气候融资 |
| carbon neutrality | 碳中和 |
| food security | 粮食安全 |
| plant diseases and insect pests | 病虫害 |
| green finance | 绿色金融 |

## Practice

### 1. Sentence Interpreting

（1）Our planet is not on a good path for continuing to sustain life, and the Earth is facing the accelerating detrimental effects of climate change, which is also referred to as global warming.

（2）Climate change is defined as a long-term change in the Earth's overall temperature with massive and permanent ramifications. Climate scientists believe that this isn't caused by the natural process of the Earth but by human activities.

（3）The Earth's atmosphere consists of gases such as oxygen, nitrogen and other gases known as greenhouse gases like carbon dioxide, nitrous oxide and methane.

（4）Moderate greenhouse effect is natural to the Earth and is a good thing to have, but a significant greenhouse effect can be catastrophic, causing a lot of warming effect to stay in our atmosphere and heat up the Earth at an accelerating rate.

（5）Mangroves contribute to the wellbeing, food security, and protection of coastal communities worldwide.

（6）全球每年饲养超过 560 亿只动物以满足食用需求，这一数字是全球人口的 8 倍，是向大气中排放废弃物的 8 倍。

（7）滥伐或焚烧森林不仅会加剧温室气体排放，还会使植物、树木和森林消失。森林和植物能够吸收二氧化碳，帮助人类应对气候变化。我们砍伐的森林越多，它们能从大气中吸收并固定的二氧化碳就越少。

（8）固体垃圾填埋场是美国最大的甲烷气体污染源之一。每个人平均每天产生 4.3 磅垃圾，一年就是 1,569 磅。地球上有 70 多亿人口，每年有 11.6 万亿磅的垃圾被倾倒到垃圾填埋场。

（9）去年，我们所处的文明世界产生了高达 400 亿吨的二氧化碳，其体积容量相当于 700 万亿立方英尺，这比 100 个珠穆朗玛峰还稍大一些。

（10）地球上最大的自然污染物是火山。如果我们对火山每年产生的碳排放量进行最大的科学估算，其排放的火山二氧化碳将达到 5 亿吨，但这还不到我们人类文明排放的近 400 亿吨二氧化碳的 2%。

## 2. English-Chinese Interpreting

### The UN Secretary-General Remarks at High-Level Opening of COP27 Climate Implementation Summit

Thank you for this wonderful hospitality and for this spectacular organization.

Your majesties,

Excellencies,

Dear friends,

In just a few days, our planet's population will cross a new threshold. The 8 billionth member of our human family will be born. This milestone puts into perspective what this climate conference is all about. What did you do for our world–and for our planet–when you had the chance?

Excellencies,

This UN Climate Change Conference is a reminder that the answer is in our hands. And the clock is ticking. We are in the fight of our lives. And we are losing.

Greenhouse gas emissions keep growing. Global temperatures keep rising. And our planet is fast approaching tipping points that will make climate chaos irreversible. We are on a highway to climate hell with our foot still on the accelerator. The war in Ukraine, other conflicts, have caused so much bloodshed and violence and have had dramatic impacts all over the world. But we cannot accept that our attention is not focused on climate change. We must, of course work together to support peace efforts and end the tremendous suffering. But climate change is on a different timeline, and on a different scale.

Today's crises cannot be an excuse for backsliding or greenwashing. If anything, they are a reason for greater urgency, stronger action and effective accountability.

Excellencies,

Human activity is the cause of the climate problem. So human action must be the solution. Action to re-establish ambition. And action to rebuild trust–especially between North and South. The science is clear: any hope of limiting temperature rise to 1.5 degrees celsius means achieving global net zero emissions by 2050. But that 1.5-degree goal is on life support–and the machines are rattling. We are getting dangerously close to the point of no return. And to avoid that dire fate, all G20 countries must accelerate their transition now–in this decade. Developed countries must take the lead. But emerging economies are also critical to bending the global emissions curve.

Last year in Glasgow, I called for coalitions of support for high-emitting emerging economies to accelerate the transition from coal towards renewables. We are making progress with the Just Energy Transition Partnerships–but much more is needed. That is why at the beginning of COP27, I am calling for a historic Pact between developed and emerging economies–a Climate Solidarity Pact. A Pact in which all countries make an extra effort to reduce emissions this decade in line with the 1.5-degree goal. A Pact in which wealthier countries and International Financial Institutions provide financial and technical assistance to help emerging economies speed their own renewable energy transition. A Pact to end dependence on fossil fuels and the building of new coal plants–phasing out coal in OECD countries by 2030 and everywhere else by 2040. A Pact that will provide universal, affordable, sustainable energy for all.

Excellencies,

We also desperately need progress on adaptation–to build resilience to the climate disruption to come. Today, some three-and-a-half billion people live in countries highly vulnerable to climate impacts. In Glasgow, developed countries promised to double adaptation support to US $40 billion a year by 2025. We need a roadmap on how this will be delivered. And we must recognize that this is only a first step. Adaptation needs are set to grow to more than US $300 billion a year by 2030. Half of all climate finance must flow to adaptation.

The good news is that we know what to do and we have the financial and technological tools to get the job done. It is time for nations to come together for implementation. It is time for international solidarity across the board. Solidarity that respects all human rights and guarantees a safe space for environmental defenders and all actors in society to contribute to our climate response. Let's not forget that the war on nature is in itself a massive violation of human rights. We need all hands on deck for faster, bolder climate action. A window of opportunity remains open, but only a narrow shaft of light remains. The global climate fight will be won or lost in this crucial decade–on our watch. One thing is certain: those that give up are sure to lose. So, let's fight together and let's win. For the 8 billion members of our human family–and for generations to come.

Thank you.

### 3. Chinese-English Interpreting

#### 气候变化带来的影响

在全球变暖、极端天气事件频发的背景下，气候变化已经上升为全球经济增长的核心挑战之一。其影响范围已超越农业和工业，对金融服务业的冲击也日益凸显，成为不可忽视的重要因素。从多重视角审视，气候变化正对全球构成前所未有的严峻挑战。它不仅成为全球经济宏观战略中不可或缺的一环，更是关乎人类未来存亡的重要议题。国际社会必须携手合作，共同应对气候变化的威胁。

粮食安全是全球经济发展的基石，然而气候变化正从多个方面对全球粮食安全造成冲击。农作物依赖特定的生长条件，如温度、降水和日照等。随着全球变暖，升温导致粮食作物的生长期缩短。在一定范围内，适度的温度上升可能增加粮食产量，但随着气温的持续升高，病虫害可能增多，作物的灌浆时间减少，这不仅影响粮食产量，还会降低粮食的营养成分和品质。同时，降水的不稳定性也对粮食产量构成威胁。干旱会直接导致农作物减产，而适度的降水增加则有助于粮食增产。然而，极端天气如洪涝等则会对粮食产量产生严重负面影响，甚至影响粮食的储存和运输等环节。

气候变化对全球能源安全构成了显著冲击。为了减缓全球变暖的步伐，各国纷纷推出减碳的气候政策，促使可再生能源的渗透率稳步提升。然而，由于可再生能源高度依赖风、光、水等自然资源，这些资源在气候变化的影响下，特别是在极端天气条件下，变得更为脆弱，进而加剧了电力系统的脆弱性。从气温角度而言，极端高温会降低电池板的发电效率并缩短其使用寿命，而极端低温则可能导致风电机组叶片结冰等故障频发。在降水方面，降水量的增加可能会减少光照时长，这对太阳能发电造成一定影响，但同时也可能增加水力发电的产能。气候变化无疑给全球经济带来了巨大挑战，但同时也孕育着新的发展机遇。综合考虑，遏制气候变化的长期收益将远超短期成本。目前，全球范围内应对气候变化的国际合作正在不断加强，我国在绿色金融国际标准、全球碳市场对接等相关机制设计方面，已积极开展了多项双边和多边国际合作。这些合作与努力将有助于推动全球经济的绿色复苏和增长。

## 4. English-Chinese Interpreting

### What is Carbon Neutrality?

In the simplest of terms, carbon neutrality means having net-zero carbon dioxide emissions. To achieve this, you have to balance carbon emissions with carbon removal. For any amount of carbon a company or country produces, it should be taking steps to remove the same amount of carbon from the atmosphere if it wants to achieve carbon neutrality.

Many companies, organizations and countries will attempt carbon offsetting, which is a term to describe creating less or no emissions in one place to make up for carbon emissions elsewhere. Unfortunately, it's not a perfect system and there's plenty of criticism about carbon offsetting being less than effective.

When it comes to achieving carbon neutrality as an individual, it can be a little tricky. Most people don't have the time, resources, or money to ensure they are removing the amount of carbon emitted into the atmosphere more than or at least the same as they are producing, but that doesn't mean you're helpless.

There are a ton of things people can do to lower their carbon footprint. Drive your car less, plant your own vegetables, use cold water when you wash your clothes to reduce energy consumption, and of course, reduce, reuse and recycle whenever possible. You can also take a look at how you spend your money on tech.

According to Apple's own product environmental report, one iPhone 11 produces about 79 kg of carbon emissions during its lifetime. Most of those emissions (79% according to the report) come from production, while the device is actually being made.

One of the biggest things you can do to reduce your carbon footprint is not buying technology frequently. It's far too easy to get a new iPhone every year or two years, but the reality is that you don't really need to. I'm not saying buying iPhones is the only problem, most purchased goods make the most carbon emissions during production, so if you use things or keep things longer, you're reducing your carbon footprint–it's something you should think about.

### 5. Chinese-English Interpreting

**人类活动所造成的温室气体**

气候变化已成为我们这个时代的鲜明标志，我们正处于一个历史的转折点。它正在全球范围内引发前所未有的影响，包括天气模式的改变，对粮食生产的威胁，以及海平面上升导致的灾难性洪灾风险加剧。如果现在不及时采取果断行动，未来适应这些影响的难度将大大增加，成本也将更加高昂。

温室气体，这种能够吸收和反射辐射、在地球表面和大气层之间形成保温层的自然物质，对维持地球温度使其适宜生物居住至关重要。然而，经过 150 多年的工业化进程、森林砍伐和大规模农业生产，大气中温室气体的含量已攀升至 300 万年来的最高点。随着人口增长、经济发展和生活水平提高，温室气体排放总量也在持续攀升。

科学已经给出了明确的结论：地球大气中温室气体的浓度与全球平均气温直接相关；自工业革命以来，温室气体浓度不断上升，全球平均气温也随之增加；其中，二氧化碳（$CO_2$）是最主要的温室气体，约占总量的三分之二，主要来源于化石燃料的燃烧。这些科学事实提醒我们，必须正视气候变化的严峻性，并立即采取行动减少温室气体排放，保护我们共同的家园。

## Related knowledge

**United Nations Legal Instruments**

**United Nations Framework Convention on Climate Change (UNFCCC)**

The UN family is at the forefront of the effort to save our planet. In 1992, its “Earth Summit” produced the United Nations Framework Convention on

Climate Change (UNFCCC) as a first step in addressing the climate change problem. Today, it has near-universal membership. The 197 countries that have ratified the Convention are Parties to the Convention. The ultimate aim of the Convention is to prevent "dangerous" human interference with the climate system.

**The Kyoto Protocol**

By 1995, countries launched negotiations to strengthen the global response to climate change, and, two years later, adopted the Kyoto Protocol. The Kyoto Protocol legally binds developed state Parties to emission reduction targets. The Protocol's first commitment period started in 2008 and ended in 2012. The second commitment period began on 1 January 2013 and ended in 2020. There are now 198 Parties to the Convention and 192 Parties to the Kyoto Protocol.

**The Paris Agreement**

At the 21st Conference of the Parties in Paris in 2015, Parties to the UNFCCC reached a landmark agreement to combat climate change and to accelerate and intensify the actions and investments needed for a sustainable low carbon future. The Paris Agreement builds upon the Convention and–for the first time–brings all nations into a common cause to undertake ambitious efforts to combat climate change and adapt to its effects, with enhanced support to assist developing countries to do so. As such, it charts a new course in the global climate effort.

The Paris Agreement's central aim is to strengthen the global response to the threat of climate change by keeping the global temperature rise this century well below 2 degrees Celsius above pre-industrial levels and to pursue efforts to limit the temperature increase even further to 1.5 degrees Celsius.

On Earth Day, 22 April 2016, 175 world leaders signed the Paris Agreement at United Nations Headquarters in New York. This was by far the largest number of countries ever to sign an international agreement on a single day. There are now nearly 200 countries that have ratified the Paris Agreement.

**The Nobel Peace Prize**

In 2007, the Nobel Peace Prize was awarded jointly to former United States Vice-President Al Gore and the IPCC "for their efforts to build up and disseminate greater knowledge about man-made climate change, and to lay the foundations for the measures that are needed to counteract such change."

## 口译理解与短时记忆训练——句子层面

### 1. 句子元素与关系的概念

短时记忆的实质是对意义的记忆，而不是词汇的记忆。就句子而言，句子的意义由元素和关系组成。元素是指能够表达实体意义的句子成分，而关系是指能够表达句子成分具有的联系与逻辑。对句子的口译训练，主要训练译者快速抓住元素与关系并能够迅速转换成译语的信息处理能力。

### 2. 句子元素与关系的识别

（1）It is true that lack of opportunity to education is one of the factors that contribute to the problems of the homeless and poverty. (24 words)

汉译：缺乏受教育的机会正是导致流浪和贫穷的因素之一。

例（1）中的句子元素有：lack of opportunity to education，the factors，the problems，the homeless and poverty。可以看出，元素主要是名词短语，表达实在的意义。

例（1）中的句子关系有：it is true that...is，contribute to，of。可以看出，关系有从句、动词、介词。

（2）Two recent economic reports warn that although the world economy is growing, complicating issues like lower oil prices and conflict could hurt its progress. (24 words)

汉译：两份最近的经济报告警告称，虽然世界经济正在增长，但油价下跌和冲突等复杂问题可能会影响这一进程。

例（2）中的句子元素有：Two recent economic reports，the world economy，

complicating issues，lower oil prices，conflict，its progress。显然，元素都是名词短语，表达实在的意义。

例（2）中的句子关系有：warns that，although，is growing，like，hurt。这些表达有动词、从句、连词、介词、动词，主要体现了各元素之间的关系，如关系有主次之分。

（3）Have you ever noticed that people who cannot read or write usually have better memories than those who can? (19 words)

汉译：你是否曾注意到不识字的人的记忆力通常比识字的人要好?

例（3）中的句子元素有：you，people，better memories，those。这些表达有名词、代词，构成该句的元素。

例（3）句子中的主要关系有：noticed that，have...than...，次要关系有：who can't read or write，who can。

从以上三个例子我们可以总结出：句子的元素主要体现在名词词组、代词等实体成分上，表达实质性意义，而句子关系主要体现在动词、连词、介词、从句等连接实词的成分上，表达逻辑与关系，且有主次之分。

### 3. 句子元素与关系的记忆训练

（1）元素的记忆训练。元素的记忆训练可以分主题进行。政治、经济、文化、教育、科技等领域的主题元素要进行积累与训练。请听下面的句子，然后回忆出你所听到的元素，注意每个句子仅听一遍。

A long time ago, in England, some judges often decided whether a criminal was telling the truth by giving him some dry bread.

汉译：很久以前，在英国，法官通过给罪犯一些干面包来判断他是否讲的是真话。

The goal is to find out whether kids using today's new technology can teach themselves to read in places where no schools or teachers exist.

汉译：我们的目标就是考察孩子能否在没有学校或教师的地方使用新技术进行自学。

The Canadian study shows that after one year of training in music, children could get more grades in a memory test than others.

汉译：加拿大的研究表明，在音乐方面接受一年训练的儿童在记忆力测试中的分数比其他孩子要更高。

Inequality between men and women results in poorer health for children and greater family poverty, according to a new study.

汉译：一份新的研究表明，男女的不平等是儿童健康较差的原因，也是家庭更贫穷的原因。

Several people have opened a controversial institute in Shanghai that is modeled after the education system of ancient Chinese family schooling to teach children.

汉译：几个人在上海创立了一个有争议的机构，该机构效仿中国古代的家庭私塾教育制度来教育学生。

（2）关系的记忆训练。关系的记忆训练需基于元素记忆训练，循序渐进。主要关系的训练有：动词、连词、从句等记忆训练。次要关系的训练有：介词记忆训练。请听下面的句子，然后回忆出你所听到的元素与关系，注意每个句子仅听一遍。

The cultural experts try to preserve the interesting old customs against the impacts of the modern world.（目的关系）

汉译：文化专家尽力保留古老的习俗，使其免受现代社会的影响。

Even though there are strict rules regulating the use of chemicals in food, some food producers disregard them intentionally in order to gain more profit.

（让步关系、目的关系）

汉译：尽管食品中化学品的使用有严格的规定，但是一些生产者为了获利仍然对其置之不理。

Culture shock is due to our own lack of understanding of other people's cultural background and our lack of the means of communication rather than the hostility of an alien environment.（因果关系）

汉译：文化冲突是由于我们缺乏对他人文化背景、交流方式的了解，而不是异域环境的不友好造成的。

The current examinations in Korea require that students memorize facts in high school classes, but don't test critical thinking or creative problem solving abilities, so officials plan to change the admission test to also measure imagination and the ability to innovate.（因果关系）

汉译：韩国目前的考试要求学生们背诵高中课程内容，但并不考察批判性思维或者创造性解决问题的能力，所以官员计划对招生考试进行改革，以增加对想象力和创新能力的考察。

The American government will invest three billion dollars in the education of young Americans in science, technology, engineering, and math in the hope of changing the situation that not enough American students want to be engineers, mathematicians, or scientists.（目的关系）

汉译：美国政府将投资 30 亿美元用于美国年轻一代在科学、技术、工程和数学方面的教育，希望能改变想成为工程师、数学家或者科学家的美国学生并不多的事实。

## 1. Sentence Interpreting

（1）我们的地球并未沿着一条能够持续维持生命的健康道路前行，而是正面临着日益加剧的气候变化所带来的严重不利影响，即全球变暖。

（2）气候变化指的是地球整体温度的长期变化，这种变化会带来广泛且持久的影响。气候科学家普遍认为，这种变化并非源于地球自身的自然过程，而是主要由人类活动所引发。

（3）地球的大气层主要由氧气和氮气等气体构成，同时包含一系列被称为温室气体的成分，如二氧化碳、一氧化二氮和甲烷。

（4）存在适度的温室效应是地球的自然现象，并且有益于地球环境。

（5）红树林在增进全球沿海社区的福祉、保障粮食安全方面发挥着重要作用。

（6）Every year, more than 56 billion animals are raised for food in the world, which is 8 times the human population and 8 times the amount of waste discharged into the atmosphere.

（7）Forest burning and deforestation not only emits greenhouse gases but also removes plants, trees and forests, which are the things on earth that absorb carbon dioxide and are helpful for dealing with climate change. The more deforestation, the less carbon dioxide the forest can absorb from the atmosphere.

（8）Solid waste landfills are one of the largest methane gas pollutants in the United States. On average, a person produces 4.3 pounds of waste per day, or 1,569 pounds per year. With more than 7 billion people on Earth, 11.6 trillion pounds of waste is dumped into landfills every year.

（9）Our civilized world produced 40 billion tons of carbon dioxide last year, which is equivalent to a volume of 700 trillion cubic feet, slightly larger than 100 Mount Qomolangma.

（10）The largest source of natural pollutant on Earth is volcanoes. If we take the largest scientific estimate of the amount of carbon emissions volcanoes produce each year, it would be 500 million tons of volcanic $CO_2$, but that's less than 2 percent of the nearly 40 billion tons of carbon dioxide our human civilization produces.

## 2. English-Chinese Interpreting

### 联合国秘书长古特雷斯在 COP27 气候执行峰会高级别开幕式上的讲话

非常感谢你们的热情款待以及精心组织了此次盛大的峰会。

陛下，诸位阁下，亲爱的朋友们！

几天后，我们星球的人口将迈过一个重要的关口——人类大家庭将迎来第 80 亿个成员。这一历史性的时刻使我们更加深刻地认识到本次气候会议的重大意义：在这个星球上，我们每一个人都应该思考，我们能为世界、为地球做些什么？

诸位阁下，

本次联合国气候大会是一次重要的提醒，答案其实就在我们每个人的手中。时间紧迫，我们实际上是在为自己的未来而战，但目前的形势并不乐观。

温室气体排放量不断攀升，全球气温持续升高。我们的星球正日益接近临界点，一旦越过，气候混乱将变得不可逆转。我们仿佛正通往气候地狱的高速公路上，油门却仍然被踩得紧紧的。尽管乌克兰战争和其他冲突带来了诸多流血和暴力事件，给全世界带来了巨大的冲击，但我们不能因此忽略气候变化这一更为紧迫的全球性挑战。我们当然需要共同努力支持和平，结束战争与苦难。但气候变化的时间线和潜在影响要求我们采取更为紧迫和有力的行动。

当前的危机不应成为我们倒退或采取“漂绿”的借口。相反，这些危机应成为我们加紧采取行动、加强问责制的催化剂。

诸位阁下，

气候问题主要源于人类活动，因此解决方案也必须从人类的行动出发。我们需要重新点燃雄心壮志，并重建各国间的信任，特别是南北国家之间的信任。科学已经明确指出，要将全球温度上升限制在 1.5 摄氏度以内，我们必须在 2050 年前实现全球净零排放。然而，排放依然在持续，1.5 摄氏度的目标不仅关乎气候变化，更是维护地球生命系统的关键。当前，我们正处于危险的边缘。为了避免这一可怕的命运，所有 G20 国家都必须在这十年内加速向可持续发展转型。发达国家应当承担起领导责任，但新兴经济体在减缓全球排放上升方面也扮演着至关重要的角色。

去年在格拉斯哥，我倡议建立一个联盟，以支持高排放新兴经济体加速从煤炭向可再生能源的转型。我们在“公正能源转型伙伴关系”方面已取得一定进展，但仍有很长的路要走。因此，在 COP27 会议开幕之际，我呼吁发达经济体与新兴经济体之间达成一项具有历史意义的协议——气候团结公约。该公约要求所有国家在十年内，基于 1.5 摄氏度的目标，做出额外的努力来减少排放。富裕国家和国际金融机构需提供金融和技术援助，帮助新兴经济体加速其可再生能源转型。我们必须结束对化石燃料的依赖，并停止建设新的燃煤电厂，包括经济合作组织国家在 2030 年前逐步淘汰煤炭，其他地区在 2040 年前逐步淘汰煤炭。这是一项旨在为全球所有人提供普遍的、可负担的且可持续的能源的协议。

诸位阁下，

我们还需迫切加强在气候适应方面的努力，以应对即将到来的气候破坏。当前，约 35 亿人生活在极易受气候影响的地区。在格拉斯哥，发达国家承诺到 2025 年将气候适应力的支持增加一倍，达到每年 400 亿美元。然而，我们仍需制定明确的路线图来实现这一目标。我们必须认识到，这仅仅是开始，到 2030 年，气候适应力的需求将增长到每年 3,000 亿美元以上。因此，我们必须确保至少一半的气候资金用于增强适应力。

好消息是，我们清楚知道应该做什么，并且拥有所需的金融和技术工具。现在是各国齐心协力、共同行动的时候了。让我们团结一致，尊重所有人权，确保环境维护者和所有社会参与者能够安全地参与气候行动。我们不应忘记，对自然的战争实质上是对人权的大规模侵犯。我们需要所有人齐心协力，采取更快、更大胆的气候行动。机会之窗仍然敞开，但时间已非常紧迫。在这关键的十年里，应对全球气候挑战的成败将取决于我们的行动。有一点是明确的：放弃的人注定会失败。因此，让我们并肩战斗，共同取得胜利。为了人类大家庭的 80 亿成员，为了我们的子孙后代。

谢谢。

### 3. Chinese-English Interpreting

## The Impact of Climate Change

Against the backdrop of global warming and increasing extreme weather events, climate change has become one of the major challenges to global economic growth. The impact of global climate change is no longer limited to agriculture and industry, and its impact on the financial services industry is becoming more evident. To some extent, climate change poses severe challenges to the world. Climate change is not only one of the keys to the global economic macro strategy, but also an important issue concerning the survival of mankind. Addressing climate change is an area where the international community must work together.

Food security is the foundation of global economic development, and climate change has impacts on global food security in many ways. Since crops have their suitable growth temperatures, precipitation, sunlight, etc., with global warming and rising temperatures, the growth period of food crops will be shortened, and the temperature rise within a certain range may increase the grain yield, but as the temperature continues to rise, plant diseases and insect pests may increase, and the filling time of crops will be reduced, which will affect grain production, and will also reduce the nutrient content of grain, resulting in a decrease in quality. In addition, precipitation volatility will also affect the food production. Drought will lead to a decrease in grain yield, while an increase in moderate precipitation will increase grain yield. However, extreme weather such as floods will have a negative impact on food production. It will also influence food storage and transportation.

Climate change also has an impact on global energy security. In order to slow down the process of global warming, countries have adopted climate policies on carbon reduction, so as to increase the penetration rate of renewable energy. However, due to the high dependence of renewables on natural resources such as wind, solar, and water, these resources are more vulnerable to climate change, especially in the face of extreme weather, thereby increasing the vulnerability of the power system. For temperature, extreme high temperatures will affect the power generation efficiency and service life of battery panels, while extreme low temperatures will cause faults such as icing of wind turbine blades. For precipitation, with the increase of precipitation, the amount of sunlight may decrease, which affects solar power generation to a certain extent, but it may increase the power generation capacity of hydropower at the same time. Although climate change poses severe challenges to the global economy, it presents new opportunities. So far, the long-term benefits of curbing climate change far outweigh the short-term costs. At present, international communities are enhancing cooperation in the field of climate change continuously. China has also carried out a number of bilateral and multilateral international cooperation in international standards for green finance and in the design of relevant mechanisms for connecting global carbon markets. These cooperation and efforts will promote the green development of the global economy.

### 4. English-Chinese Interpreting

**什么是碳中和？**

碳中和，简而言之，就是实现二氧化碳的净零排放，这要求我们必须平衡碳排放与碳吸收。对于企业或国家而言，要想实现碳中和，就需要采取相应措施，确保从大气中移除的碳量与排放的碳量相等。

众多企业、组织和国家会尝试采用碳抵消策略，即通过减少某处的碳排放来补偿其他地方的排放。然而，这一体系并不完美，其效果也饱受批评。

对于个人而言，实现碳中和可能较为棘手。大多数人难以投入足够的时间、资源或资金来确保个人的碳移除量大于或等于排放量。但这并不意味着我们无法为碳中和做出贡献。

实际上，每个人都可以从日常生活中着手，减少自己的碳足迹。例如，减少驾车出行，自己种植蔬菜，使用冷水洗衣以减少能源消耗，以及尽可能地减少使用、重复使用和回收物品。不妨看一下你在科技产品方面的消费（科技产品的消费也是碳排放的一个重要来源）。

据报告，以苹果公司的 iPhone 11 为例，其生命周期内的碳排放量约为 79 千克，其中大部分（约 79%）来自设备的生产环节。

因此，减少碳足迹的一个有效方法就是减少科技产品的购买频率。虽然每年或每两年更换一部新 iPhone 似乎很方便，但实际上并非必要。不仅 iPhone 如此，大多数商品在生产过程中的碳排放都是最高的。因此，延长科技产品的使用寿命，可以有效减少个人的碳足迹，这是一个值得我们深思的问题。

**5. Chinese-English Interpreting**

**The Human Fingerprint on Greenhouse Gases**

Climate Change is the defining issue of our time and we are at a defining moment. From shifting weather patterns that threaten food production, to rising sea levels that increase the risk of catastrophic flooding, the impacts of climate change are global in scope and unprecedented in scale. Without drastic action today, adapting to these impacts in the future will be more difficult and costly.

Greenhouse gases occur naturally and are essential to the survival of humans and millions of other living things, by keeping some of the sun's warmth from reflecting back into space and making Earth livable. But after more than a century and a half of industrialization, deforestation, and large scale agriculture, quantities of greenhouse gases in the atmosphere have risen to record levels not seen in three million years. As populations, economies and standards of living grow, so does the cumulative level of greenhouse gas (GHGs) emissions.

There are some basic well-established scientific links: The concentration of GHGs in the earth's atmosphere is directly linked to the average global temperature on Earth; The concentration has been rising steadily, and global mean temperatures along with it, since the time of the Industrial Revolution; The most abundant GHGs, accounting for about two-thirds of GHGs, carbon dioxide ($CO_2$), is largely the product of burning fossil fuels. These scientific facts serve as a reminder that we must acknowledge the gravity of climate change and take prompt action to mitigate greenhouse gas emissions and safeguard our shared planet.

# Unit 4 Biodiversity Conservation 生物多样性保护

## Introduction

### What is Biodiversity?

The term biodiversity (from "biological diversity") refers to the variety of life on Earth at all its levels, from genes to ecosystems, and can encompass the evolutionary, ecological, and cultural processes that sustain life. Biodiversity includes not only species we consider rare, threatened, or endangered but also every living thing–from humans to organisms we know little about, such as microbes, fungi, and invertebrates.

### Why is Biodiversity Important?

Biodiversity is important to most aspects of our lives. We value biodiversity for many reasons, some utilitarian, some intrinsic. This means we value biodiversity both for what it provides to humans, and for the value it has in its own right. Utilitarian values include the many basic needs humans obtain from biodiversity such as food, fuel, shelter, and medicine. Further, ecosystems provide crucial services such as pollination, seed dispersal, climate regulation, water purification, nutrient cycling, and control of agricultural pests. Biodiversity also holds value for potential benefits not yet recognized, such as new medicines and other possible unknown services. Biodiversity has cultural value to humans as well, for spiritual or religious reasons for instance. The intrinsic value of biodiversity refers to its inherent worth, which is independent of its value to anyone or anything else. This is more of a philosophical concept, which can be thought as the inalienable right to exist. Finally, the value of biodiversity can also be understood through the lens of the relationships we form and strive for with each other and the rest of nature. We may value biodiversity because of how it shapes who we are, our relationships to each other, and social norms. These relational values are part of peoples' individual or collective sense of wellbeing, responsibility for, and connection with the environment. The different values placed on biodiversity are important because they can influence the conservation decisions people make every day.

## Glossary

| | |
|---|---|
| climate regulation | 气候调控 |
| seed dispersal | 种子传播 |
| pollination | 授粉 |
| water purification | 水体净化 |
| nutrient cycling | 养分循环 |
| biodiversity | 生物多样性 |
| species | 种、物种 |

| | |
|---|---|
| conservation | 保护、保育、保存 |
| population | 种群 |
| biodiversity hotspot | 生物多样性热点区 |
| poach | 偷猎 |
| national park | 国家公园 |
| logging | 伐木 |
| Convention on Biological Diversity (CBD) | 《生物多样性公约》 |
| biodiversity loss | 生物多样性丧失 |
| ecological degradation | 生态系统退化 |
| sustainable development | 可持续、绿色发展 |
| taxon | 类群 |
| flagship species | 代表性物种，旗舰物种 |
| endangered species | 濒危物种 |
| vertebrate | 脊椎动物 |
| gene bank of species | 物种基因库 |
| Yunnan snub-nosed monkey | 滇金丝猴 |
| habitat destruction | 生境破坏 |
| higher plant | 高等植物 |
| Ma | 百万年 |
| domesticated | 家养、驯养 |
| cultivated | 养殖的、培育的 |

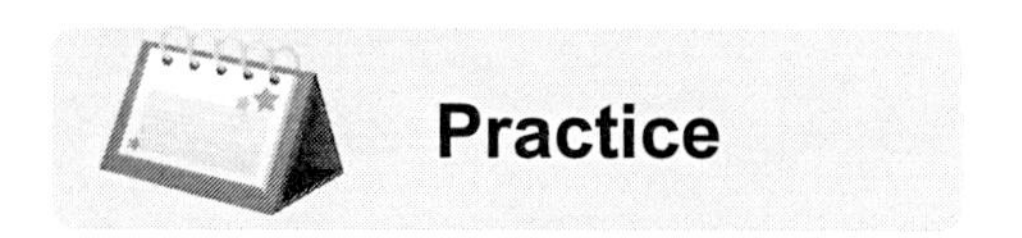

## Practice

### 1. Sentence Interpreting

（1）Although Yunnan accounts for only 4.1% of China's land area, the province is deemed Kingdom of Animals and Plants and Gene Bank of Species because it harbors more than 50% of the country's biological species.

（2）In recent years, China has made significant achievements in ecological conservation with the enforcement of laws, policies and other measures, such as the construction of national parks and a ban on forest logging.

（3）Yunnan has established 166 nature reserves at different levels, protecting about 90% of typical ecosystems and 85% of important species, including Asian elephants and Yunnan snub-nosed monkeys.

（4）In 2020, several government departments teamed up to ameliorate problems such as the destruction of coral reefs, illegal mining of sea sand, illegal dumping of waste, and human reclamation of wetlands.

（5）With a 4,060 km border, Yunnan has also enhanced cross-border biodiversity conservation with countries including Laos, Myanmar and Viet Nam.

（6）多边机构和政府应采纳土著人民拥有的传统知识、独特做法和创新策略，以促进生物多样性保护。

（7）土著人民凭借其丰富的知识和持续的创新，不仅保护着自然生态，还为实现气候目标以及增强社区在疫情中的适应能力做出了重要贡献。

（8）几个世纪以来，土著社区在保护生物多样性热点区域表现出了非凡的才智，这些热点区域是全球粮食、能源和水安全的基础。

（9）长江长达 6,300 千米，其流域地势复杂、气候多样，是世界上生物多样性最丰富的区域之一。

（10）我们需要一个新的全球生物多样性框架，以解决生物多样性丧失的驱动因素。

## 2. English-Chinese Interpreting

### The Giant Panda Conservation

The giant panda was one of the most endangered mammals on the earth, a flagship species for wildlife conservation. Its evolutionary history can be traced back 7-8 Ma to the late Miocene and the earliest recorded ancestor. Because of human-induced habitat loss and fragmentation, the giant panda remains restricted to six isolated mountain ranges across China. In fact, human activities had deeply affected this species and almost brought them to the brink of extinction. Research indicated the endangerment of giant pandas was closely related to environmental pressures brought by growing human populations throughout the Chinese history. Deforestation, road and settlement construction, land reclamation, mineral exploitation, and large water engineering projects and even poaching were the primary risk factors to pandas: they caused habitat loss and fragmentation.

To solve these problems, the Chinese government has implemented laws and regulations over the past decades. In 1988, the Chinese Government enacted the Law on the Protection of Wildlife to protect endangered animals from human disturbance. Since then, poaching has been banned and poachers have been severely punished. National conservation measures for giant pandas or closely related national initiatives have been taken to save giant pandas, including the ban of poaching, the panda habitat protection project, the panda nature reserve network, the natural forest protection project, and the Grain for Green project.

These endeavors have led to significant conservation achievements: poaching has been eradicated; 67 giant panda nature reserves have been established, protecting approximately 54% of panda habitats with over 66% of individuals; the fourth national survey showed a population increase and habitat expansion. Recently, the Giant Panda National Park has been established and this will take the giant panda conservation to a new level.

### 3. Chinese-English Interpreting

中国是全球生物多样性极为丰富的 12 个国家之一，统计数据显示，其生物多样性位居世界第八，北半球之首。然而，随着庞大的人口压力和经济的迅猛发展，对资源的需求和对环境产生的影响，中国也成了生物多样性面临最严峻威胁的国家之一。由于生态系统遭受大面积破坏和退化，中国的众多物种已被列为濒危物种或受威胁物种。其中，高等植物濒危物种高达 4,000 ~ 5,000 种，占中国高等植物总种数的 15% ~ 20%。

云南，作为中国野生生物种类资源和生态系统类型最为丰富的地区，是全球公认的生物多样性关键地区之一。云南因其独特的地理位置和复杂多变的自然地理环境，形成了复杂的生态系统和多样化的生境条件，从而孕育了极为丰富的生物资源。尽管云南省的土地面积仅占全国总面积的 4.1%，但其在生物类群上却占据了举足轻重的地位。云南拥有高等植物 16,577 种，占全国总数的近 50%；脊椎动物 1,836 种，占全国总数的 55.35%。在中国公布的 335 种重点保护野生动物中，云南独占 243 种，占比高达 72.5%，其中约 20%的物种为云南所特有或在国内仅见于云南。云南的生物丰富度、特有度和特有率均居全国之首，被誉为“植物王国”和“动物王国”。因此，深入认识和研究云南生物多样性，对中国和全球的生物多样性保护与利用都具有极为重要的意义。

### 4. Chinese-English Interpreting

#### 认识亚洲象

亚洲象是热带森林生态系统的标志性物种，被列为我国一级重点保护野生动物，被世界自然保护联盟（IUCN）列为濒危（EN）等级物种。历史上，野生亚洲象分布区域很广，目前仅分布于印度、斯里兰卡、泰国、越南、柬埔寨、老挝、马来西亚、缅甸、孟加拉国、不丹、印度尼西亚、尼泊尔及中国等 13 个国家和地区。全球估计有 4.1 万 ~ 5.2 万头亚洲象。在我国，野生亚洲象主要栖息于云南。根据 2021 年调查统计，云南省约有亚洲象 300 多头，数量远低于中国的大熊猫。

在中国文字中，“象”与“祥”同音，大象被人们视为吉祥、富足和力量的象征。史书中还有很多大象帮助人类祖先耕种、击退敌军的故事。亚洲象是亚洲现存的最大的陆生动物，以家族为单位群居。一般由年龄最长、经验最丰富的雌象为首领。亚洲象拥有超群的记忆力，能记住远处的觅食地点、水源地和迁移路线。鼻子是大象的重要器官，除了嗅觉等感观功能，还能吸水、喷淋、抓取食物，也能通过握鼻或用鼻子抚摸对方身体表达问候、安慰等高级情感。亚洲象是植食性动物，每天要进食 10 多个小时，总进食量可达 150 千克。

随着《中华人民共和国野生动物保护法》等法律法规的深入实施，我国野生亚洲象的数量持续增长。自 20 世纪 90 年代起，亚洲象开始季节性地走出保护区进入农田觅食，其活动范围也随之不断扩大。迁移是野生亚洲象的自然习性，通过种群迁移扩散，它们能够寻找新的水源、食物和栖息地，促进种群间的基因交流。

## 5. English-Chinese Interpreting

### What are Genetic Resources and Why are They Important?

All living organisms, plants, animals and microbes, carry genetic materials that could be potentially useful to humans. These resources can be taken from the wild, domesticated or cultivated. They are sourced from environments in which they occur naturally, or from human-made collections such as botanical gardens, gene banks, seed banks and microbial culture collections.

There are significant potential benefits to be gained by accessing genetic resources and making use of them. They provide a crucial source of information to better understand the natural world and can be used to develop a wide range of products and services for human benefit. This includes products such as medicines and cosmetics, as well as agricultural and environmental practices and techniques. However, like many key resources in the world, genetic resources are not evenly distributed. What's more, the plants, animals and microbes in which they are found often make up complex and delicately-balanced ecosystems which can be threatened or endangered. The way in which genetic resources are accessed, and how the benefits of their use are shared, can create incentives for their conservation and sustainable use, and can contribute to the creation of a fairer and more equitable economy to support sustainable development.

Our current understanding of genetic resources owes a great deal to the traditional knowledge of indigenous and local communities. This valuable knowledge has been built up and handed down over generations. It is essential that the value of traditional knowledge is understood and valued appropriately by those who use it, and that the rights of indigenous and local communities are considered during negotiations over access and use of genetic resources. Failure to do this can put the knowledge, the resources and the communities at risk.

## Related knowledge

### (1) The logo of UN CBD COP 15

**One Planet**

The Earth is the only planet human beings have, and is the common home to all life.

**One Water Drop**

Water is the source of life.

**One Community of Life**

All living beings on Earth are inseparable, depending on each other for life and death, rise and fall.

**One Totem Full of Cultural Elements of the Host Country**

The oracle bone inscriptions, Chinese seal and paper-cut embody the long-standing Chinese culture; the most treasured giant panda is a messenger of friendship; the graceful peacock prays for auspiciousness; the plum blossoms stand for the confidence, determination and courage, with which we all, in the face of present severe situation of conserving global biodiversity, strive to realize the vision of Living in Harmony with Nature.

**One Seed**

A seed carries vitality, vigor and hope for new life. International community expects with full confidence that the outcomes of CBD COP15 will nourish all living creatures and benefit our future generations.

**(2) Conservation *vs.* Preservation**

Earth's natural resources include air, minerals, plants, soil, water, and wildlife. Conservation is the care and protection of these resources so that they can persist for future generations. It includes maintaining diversity of species, genes, and ecosystems, as well as functions of the environment, such as nutrient cycling. Conservation is similar to preservation, but while both relate to the protection of nature, they strive to accomplish this task in different ways. Conservation seeks the sustainable use of nature by humans, for activities such as hunting, logging, or mining, while preservation means protecting nature from human use.

This difference is illustrated by how the United States manages its public lands. The goal of National Parks, for instance, is preservation with an emphasis on causing minimal change to the landscape or environment, meanwhile National Forests can be used for cattle grazing, logging, hunting, and recreation.

## 口译理解与短时记忆训练——段落层面

段落层面的短时记忆训练建立在句子的基础之上，同时也具有段落的特点。段落的短时记忆，需要从以下方面进行：

（1）抓住段落主题及关键词。

（2）抓住段落结构。

（3）抓住支撑信息、过渡词及内部逻辑。

### 1. 抓住段落主题及关键词

段落主题就是对段落大意的概括与总结。它有时直接出现在文章的开始、末尾或者中间，有时不会直接出现，需要口译员主动听取信息进行加工处理，进入短时记忆，以帮助他们更好地完成口译任务。关键词就是能够帮助口译员总结、概括、提炼段落大意的词语，善于快速、准确捕捉关键词对于口译员提高口译段落的记忆质量具有重要的作用。抓住段落主题及关键词是训练口译员段落短时记忆的第一步。

例（1）：请用一句话概括以下段落主题，并找出关键词。

To get a driver's license, you must take a series of tests. The first test is the vision test. The second test is a written test. The last test you have to take is a road test. If you pass them, you will receive your driver's license.

分析：第一句话就是主题：To get a driver's license, you must take a series of tests。其中的关键词是：driver's license，tests。抓住了这些信息，我们就听懂了大意，记住了这个段落到底在说什么。这个例子直接给出了主题，不需要口译员去总结。

例（2）：请用一句话概括以下段落主题，并找出关键词。

Most students rely on their parents to pay the tuition fee. Students can apply for a scholarship. They can also apply for low-interest loans. In addition to the above, taking part-time jobs is also a common way for college students to get money to pay part of their tuition fees.

分析：该段落没有用直接话语来概括主题，此时就需要口译员自己主动听取信息，对所听的内容进行概括与总结。我们发现，该段讲了四种支付学费的方式。于是可以将主题总结为：there are four ways to pay the students' tuition fees。其中的关键词是：tuition fee，parents，scholarship，low-interest loan，and part-time job。抓住了这些关键信息，我们就能总结出大意，记住了这个段落的主题。这个例子没有直接给出主题，需要口译员总结，主动听取信息并进行加工处理。

## 2. 抓住段落结构

段落结构就是段落的脉络、段落的地图。寻着脉络，我们就能找到细节；沿着地图，我们就能找到更小的地方。通常，段落结构可以分为总分、分总、总分总三种结构。辨别段落结构是口译员短时记忆训练必不可少的环节之一。

例（1）：辨别以下段落结构。

Nowadays, university students can have many channels to handle their tuition fees and other expenses. Most students rely on their parents to pay the tuition fees and pay for their books, accommodation and other costs. Student can also apply for a scholarship. Most universities set up scholarships to reward students who do well in entrance exams and reward those who work well at university. They can also apply for low-interest loans. In addition to the above, taking part-time jobs is also a common way for college students to get money to pay part of their tuition fees. In a word, there are many ways for students to finance their college education.

分析：该段落是总分总的结构：channels–parents+scholarships+low-interest loans+part-time jobs–ways。这个结构可以使口译员从宏观到微观再到宏观掌握整个段落的内部脉络，更好地为主题与支撑句之间的连接服务。

例（2）：听下面一段话，然后辨别其段落结构。

We can benefit much from making our cities greener by planting trees. First, trees absorb carbon dioxide to produce oxygen which is essential to the health of human beings. Second, trees and other greenery make our cities more beautiful. Third, trees can improve the climate of the cities.

分析：听完之后，我们发现该段落是总分的结构：benefits of planting trees–health of human beings + more beautiful cities + climate of the cities。这个从主题句到支撑句的总分结构，可以使口译员掌握整个段落的脉络，从而增强其短时记忆的效果。

### 3. 抓住支撑信息、过渡词及内部逻辑

支撑信息就是进一步解释或者说明主题的信息，是段落细节的重要组成部分。它通常体现为事例、理由、事实、故事等比较具体的内容，如缺乏支撑信息的短时记忆训练就会削弱段落口译的质量与效果。不同的支撑信息通常会从不同的角度对主题进行论证，口译员需要辨别各个角度。同时，也要注意各个支撑信息之间的过渡以及支撑信息内部的过渡与连接逻辑。

例（1）：请找出以下段落的支撑信息，并指出它们之间的过渡以及其内部逻辑。

To get a driver's license, you must take a series of tests. The first test is the vision test. This test checks your eyes to see if you have to wear glasses when you drive. The second test is a written test that checks your knowledge of highway and traffic regulations. You can prepare for this test by studying a booklet that you can get at the driver's license office. After you pass the written test, you receive an instruction permit. This permit allows you to practice driving with an experienced driver. The last test you have to take is a road test. This checks your driving ability. If you pass it, you will receive your driver's license.

分析：我们已经分析过该段落的主题及关键词。接着，我们进一步分析其支撑信息与逻辑过渡。通过分析，可发现支撑信息之间的过渡是通过顺序词语实现的，the first test–the second test–after… instruction permit–the last test. 显然，该段的支撑信息是通过事物发展的顺序连接在一起的。接着，我们看各个支撑信息的内部逻辑与联系。第一个支撑信息 vision test，作者补充了该测试的目的与内容。第二个支撑信息 written test，作者补充了该测试的内容以及通过的方式。第三个支撑信息 instruction permit，作者解释了该信息的具体内容。第四个支撑信息 road test，作者解释了该测试的内容。四个信息的内部逻辑与联系主要按照测试的先后顺序进行补充说明与解释实现。通过这样的分析，口译员的段落口译能力会得到切实的训练。

例（2）请找出以下段落的支撑信息，并指出它们之间的过渡以及其内部逻辑。

To be successful in a job interview, you should demonstrate certain personal and professional qualities. First of all, you should be properly dressed. The right clothes worn at the right time can win respect of the interviewer and his confidence in your judgment. It may not be true that clothes make the man, but the first and often last impression of you is determined by the clothes you wear. Secondly, you should pay close attention to your manner of speaking, since speech is a reflection of personality. You should reflect confidence by speaking in a clear voice, loud enough to be heard, without being aggressive and overpowering. You should be prepared to talk knowledgeably about the requirements of the position you are applying for. Finally, to be really impressive you must convey a sense of self-confidence and enthusiasm for work. If you display these characteristics, with just a little luck, you'll certainly succeed in the typical personnel interview.

分析：第一个支撑信息 properly dressed，第二个支撑信息 manner of speaking，第三个支撑信息 self-confidence and enthusiasm，它们之间通过顺序词语实现了过渡：first of all–secondly–finally。通过进一步分析，我们发现三个支撑信息当中，第一个和第二个较长，具有一定的内部过渡。我们先来分析第一个支撑，很明显作者论述了衣服对于面试的重要性，能够赢得尊重与信任，能够获得第一印象，同时说明不能以"衣服"取人。可以看出，该信息通过内部的语义关系及逻辑联系阐释了该观点。第二个支撑信息，since 表明了一定程度的因果关系，同时进一步采用并列的结构论述了具体的讲话方式。这样，支撑信息的内容逻辑与过渡经过口译员的信息处理与加工实现短时记忆。

例（3）：请找出以下内容的支撑信息，并指出它们之间的过渡以及其内部逻辑。

Firms always need statistics. They usually need to calculate the average wage of their employees. In addition, companies sometimes need to determine statistically whether consumers like a new product.

分析：该段落主题与支撑信息之间存在因果关系，使用两条支撑信息作为原因。两个信息之间具有并列关系，它通过"in addition"实现过渡。可通过两条支撑信息的关键词"wage""employees""customers""like""product"实现短时记忆。

例（4）：请找出以下内容的支撑信息，并指出它们之间的过渡以及其内部逻辑。

It is unlikely that a personal assistant will be successful in her career without a knowledge of English. She may need it for translating business letters. She will probably need it as well for arranging appointments with English-speaking clients.

分析：该段落主题与支撑信息之间存在因果关系，使用两条支撑信息作为原因。两个信息之间具有并列关系，它通过“as well”实现过渡。

## 1. Sentence Interpreting

（1）云南，仅占中国国土的4.1%，却汇聚了超过全国50%的生物物种，被誉为“动植物王国”和“物种基因库”。

（2）近年来，通过实施法律、政策以及多项措施，如建设国家公园和禁止森林砍伐，中国在生态保护领域取得了显著进步。

（3）云南地区已设立166个各级各类的自然保护区，有效保护了约90%的典型生态系统和85%的重要物种，其中不乏亚洲象、滇金丝猴等珍稀物种。

（4）2020年，多个政府部门联合行动，解决了珊瑚礁破坏、海砂非法开采、垃圾非法倾倒以及人为开垦湿地等问题。

（5）云南省与老挝、缅甸、越南等邻国拥有长达4,060千米的边境线。云南与这些国家加强了跨境生物多样性保护工作。

（6）Multilateral institutions and governments should leverage the traditional knowledge, practices and innovations held by indigenous peoples to promote biodiversity conservation.

（7）The knowledge, innovations and best practices of native peoples can be leveraged to advance nature conservation, achieve climate targets and build social resilience amid pandemics.

（8）For centuries, indigenous communities have demonstrated extraordinary ingenuity in protecting biodiversity hotspots that underpin global food, energy and water security.

（9）The Yangtze River stretches 6,300 kilometers long, and its basin has complex topography and climate, making it one of the regions with the richest biodiversity in the world.

（10）We need a new global biodiversity framework that addresses the drivers of biodiversity loss.

## 2. English-Chinese Interpreting

### 大熊猫保护

大熊猫，作为地球上曾一度最为濒危的哺乳动物之一，是野生动物保护领域的旗舰物种。其进化历程可追溯到七八百万年前的中新世晚期，是地球上最早被记录的祖先。然而，由于人类活动导致的栖息地丧失和生境破碎化，大熊猫目前仅分布于中国六个偏远的山脉之中。人类活动几乎让这一物种走向灭绝的边缘。研究显示，大熊猫的濒危状态与中国历史上的人口增长带来的环境压力紧密相关。森林砍伐、道路和房屋建设、土地开垦、矿产开采、大型水利工程项目以及偷猎等人类活动，都是导致大熊猫栖息地丧失和生境破碎化的主要原因。

为了扭转这一局面，中国政府在过去几十年里颁布了相关的法律法规，例如《自然保护区管理条例》。自 1988 年起，《野生动物保护法》的颁布为濒危动物免受人类的干扰和迫害提供了法律层面的保护。偷猎行为被严厉打击，违法者受到严惩。同时，国家还推出了多项保护大熊猫的举措和倡议，包括禁止偷猎、大熊猫栖息地保护项目、大熊猫自然保护区网络建设、天然林保护工程和退耕还林项目等。

这些努力取得了显著成效：偷猎行为已得到有效遏制；全国范围内建立了 67 个大熊猫自然保护区，保护了约 54%的大熊猫栖息地及超过 66%的熊猫个体。第四次全国大熊猫调查结果显示，大熊猫种群数量有所增长，栖息地有所扩大。近期，大熊猫国家公园的成立更是标志着大熊猫保护工作迈上了一个新台阶。

### 3. Chinese-English Interpreting

China is one of the 12 countries (Mega-diversity countries) rich in biodiversity in the world. According to statistics, China's biodiversity ranks the eighth in the world and the first in the northern hemisphere. However, with huge population pressure and rapid economic development, China is among the countries where biodiversity is seriously threatened. Due to extensive destruction and degradation of ecosystems, many species in China have become threatened and quite a few endangered. In terms of higher plants, endangered species amount to 4,000-5,000, accounting for 15%-20% of the total distribution of higher plants in China.

Yunnan is home to the richest wild species resources and ecosystem types in China, and it is recognized as one of the key biodiversity areas in the world with the most concentrated distribution of important biodiversity groups. Due to Yunnan's special geographical location and its complex and unique natural geographical environment, a complex ecosystem and diverse habitat condition and extremely rich biological resources have been formed. The land area of Yunnan Province accounts for 4.1% of China's total. As for the known biological groups, Yunnan has 16,577 species of higher plants, accounting for nearly 50% of China's total. There are 1,836 species of vertebrates in the province, accounting for 55.35% of China's total; among the 335 key protected wild animals released by China, there are 243 species in Yunnan, accounting for 72.5% of China's total, and of which 20% are endemic to or only found in Yunnan of China. Its biological richness value, endemism value and endemism rate all rank the first in China, and it is known as the Kingdom of Plant and

Kingdom of Animal. Therefore, in-depth understanding and research on the biodiversity of Yunnan is necessary for the conservation and utilization of biodiversity in China and the world.

**4. Chinese- English Interpreting**

**Some Basics of Asian Elephants**

Asian elephants are the flagship species of tropical forest ecosystems, listed as China's first-class key protected wild animals, and listed as endangered (EN) species by the International Union for Conservation of Nature (IUCN). Wild Asian elephants have been widely distributed in history, but are currently only distributed in 13 countries and regions including India, Sri Lanka, Thailand, Vietnam, Cambodia, Laos, Malaysia, Myanmar, Bangladesh, Bhutan, Indonesia, Nepal and China. It is estimated that there are about 41,000 to 52,000 Asian elephants in the world. Wild Asian elephants in China are mainly distributed in Yunnan. According to the 2021 survey and statistics, there are more than 300 Asian elephants in Yunnan Province, which is far lower than the number of giant pandas in China.

In Chinese characters, "elephant" and "Xiang" (good luck) have the same pronunciation. Elephants are regarded as a symbol of auspiciousness, wealth and strength. There are also many stories in history books about elephants helping human ancestors to farm and repel enemy troops. Asian elephants are the largest existing terrestrial animals in Asia and live in groups in families. Generally, the oldest and most experienced female elephant is the leader. Asian elephants have a superb memory and can remember distant foraging locations, water sources and migration routes. The trunk is an important organ of an elephant. In addition to sensory functions such as smell, it can also absorb water, spray, and grab food. It can also express high-level emotions such as greetings and comfort by holding the trunk or touching each other's bodies with the trunk. Asian elephants are herbivorous animals, and they eat for more than 10 hours a day, with a total food intake of up to 150 kilograms.

With the implementation of laws and regulations on the protection of wild animals such as the Wild Animal Protection Law of the People's Republic of China, the number of wild Asian elephants in China has continued to increase. Since the 1990s, Asian elephants have begun to seasonally leave the protected area and to enter farmland for food. The scope keeps expanding. Migration is the living habit of wild Asian elephants, and population migration and dispersal help them find new water sources, food and habitats, and carry out gene exchange between populations.

## 5. English-Chinese Interpreting

### 什么是遗传资源，它们为什么重要？

遗传资源，即生物体（植物、动物、微生物）所携带的、对人类具有潜在价值的遗传物质，它们来源广泛，不仅存在于野生生物种中，也存在于经过驯化或培育的生物体中，甚至被精心保存在植物园、基因库、种子库和微生物菌种库等人工环境中。

遗传资源的获取与利用，不仅能为人类带来深远的利益，如促进对自然界的深入理解，并推动制药、化妆品等产业的创新，以及推进农业与环境实践和技术的发展。然而，与世界上许多重要资源一样，它们在全球范围内的分布并不均衡，更重要的是，这些资源所依托的生物体常常构成复杂而微妙的、平衡的生态系统，这些系统正面临威胁甚至崩溃的风险。遗传资源的获取方式和惠益分享机制，对于促进资源的保护与可持续利用至关重要。这有助于推动建立更公平、更公正的经济体系，以支持全球的可持续发展。

目前我们对遗传资源的深入了解，在一定程度上依赖于土著和地方社区的传统知识。这些宝贵的知识经过世代相传，在遗传资源的获取与利用过程中，其价值应得到充分的尊重与重视。在遗传资源的获取与利用的谈判中，必须充分考虑并回报/馈这些社区的权利，否则可能给传统知识、生物资源和相关社区带来不可估量的风险。

# Unit 5 Wilderness and National Parks
# 荒野与国家公园

### Wilderness

Wilderness is in the arid deserts, cypress swamps, alpine meadows, sandy beaches, and rocky crags. Wilderness protects some of the most diverse and sensitive habitats in the world. It offers refuge for wildlife and a place to seek relaxation, adventure, or something in-between for us. Although some people see a forest as wilderness, the definition of federal wilderness is specific.

The Wilderness Act of 1964 defines it as "an area where the earth and its community of life are untrammeled by man, where man himself is a visitor who does not remain." The Act's purpose is to preserve and protect the natural ecosystems and wild areas and also provide opportunities for solitude and retrospective or primitive recreation. Wilderness areas are valuable for the historical, scientific, educational, geological and ecological benefits. They help the environment and the economy.

But perhaps one of the greatest benefits is what wilderness areas can do for a person. For those who travel into wilderness areas the experience can be awe inspiring and life-changing. Those treks are what made great novels and critically acclaimed essays. Many who venture onto wilderness areas come back out changed with a deeper understanding of why these lands are set aside. Each person has their own story.

Today, there are 765 wilderness areas covering more than 109 million acres that are part of the National Wilderness Preservation System, which is managed by the Forest Service and three other federal land management services. From the 5.5-acre Pelican Island Wilderness in Northern Florida to the 9 million-acre Wrangell-Saint Elias Wilderness in Alaska, each helps shape personal stories.

## Glossary

| | |
|---|---|
| wilderness | 荒野、荒地、原野 |
| Wild Foundation | 荒野基金 |
| Forest Service | 林务局（美国） |
| sapling | 树苗 |
| national park | 国家公园 |
| delineate | 划定 |
| designate | 指定 |
| Florida Keys | 佛罗里达群岛 |
| pilot park | 试点公园 |

| | |
|---|---|
| biome | 生物群落 |
| Yellowstone National Park | 黄石国家公园 |
| Yosemite National Park | 约塞米蒂/优胜美地国家公园 |
| Three Parallel Rivers World Natural Heritage | “三江并流”世界自然遗产 |
| Shangri-La | 香格里拉 |
| sub-temperate climate | 亚温带气候 |
| arboretum | 植物园 |
| nature reserve | 自然保护区 |
| virgin/primitive forest | 原始森林 |
| fault cliff | 断崖层 |

## Practice

### 1. Sentence Interpreting

（1）National parks of the United States host millions of people a year, sustain countless species, and safeguard some of the most breathtaking landscapes on the planet.

（2）National parks preserve natural and cultural resources for the enjoyment, education, and inspiration of this and future generations.

（3）From the red sandstone formations of Utah to the Blue Ridge Mountains of Virginia, the United States holds a wealth of dazzling, resource-rich wild spaces.

（4）The government has been designating and extending protections for its public lands for over a hundred years since the U.S. Forest Service was established in 1905.

（5）The United States has added dozens of national parks to better safeguard these precious natural resources and protect valuable natural heritage for future generations.

（6）国家公园旨在保护一个或多个典型生态系统的完整性，为生态旅游、科学研究和环境教育提供场所，是需特殊保护、管理和利用的自然区域。它既非严格的自然保护区，也非一般旅游景区。

（7）国家公园作为保护区的一种形式，源自美国，如今已被世界众多国家和地区广泛采用。

（8）1872 年，美国国会正式批准设立黄石国家公园，这不仅是美国首个，也是全球最早的国家公园。

（9）追溯至 1860 年，优胜美地地区被划为美国第一座州立公园，这标志着美国国家公园历史的开端。

（10）在 2013 年 11 月党的十八届三中全会上，我国首次做出建立国家公园体制的重要决策。

## 2. English-Chinese Interpreting

A wilderness is an area of land that has been largely undisturbed by modern human development. Wilderness areas usually lack roads, buildings, and other artificial structures. They provide a natural environment for plant and animal species, and allow scientists to study healthy ecosystems.

Very few places on Earth are complete, pristine, or wildernesses. A wilderness can be reclaimed or restored, and the way a wilderness is managed can change at any time.

There are wilderness areas in every ecosystem, on every continent. The Great Barrier Reef in Australia, the world's largest coral reef, is an aquatic wilderness that is home to dozens of endangered birds, fish, mammals, and reptiles. Wilderness differs in its ranges. Some wilderness areas are huge. Most of the continent of Antarctica is one big wilderness. Some wilderness areas are small and often located in unexpected places. Ernest E. Debs Regional Park is a wilderness area in the large, densely populated urban center of Los Angeles, California, the United States.

The WILD Foundation, an organization dedicated to preserving wilderness areas around the world, defines a wilderness in two ways. First, wilderness means the natural biodiversity of a place exists in some form. Second, wilderness is also legally protected, meaning a government has laws to limit people from developing the wilderness.

People have lived with nature for thousands of years. Because of human population growth and development, many wilderness areas are threatened. The effects of human interaction with wilderness areas can be direct, such as campfire or poaching. The effects can also be indirect, such as human contribution to climate change. Climate change can limit native plant growth or lead to the growth of non-native species. Nearby development may also cause water and air pollution, threaten wilderness ecology, and even hinder animal migration and reduce biodiversity.

In order to protect wilderness, some governments limit peoples' interaction with the area. The governments also limit the kinds of human activities inside a wilderness area, placing restrictions on mining, logging, or even traveling by car.

There are two different ways to manage a wilderness area. The first is conservation, which encourages sustainable use of natural resources. Many wilderness areas have limited access to logging, for instance. This could mean a timber company can cut down a small number of trees each year if they replace the trees with saplings (young trees). It could also mean opening up parts of wilderness area to individuals for chopping down Christmas trees. The other type of wilderness management is preservation, which encourages people to preserve the wilderness by not using natural resources.

### 3. Chinese-English Interpreting

国家公园是展现一个国家自然和文化遗产以及野生动物栖息地的最具标志性的代表。中国正计划在西藏自治区设立五个被誉为“地球第三极”的国家公园。

这些“地球第三极”国家公园的核心使命在于环境保护，同时推动青藏高原的生态旅游发展，提升当地民众的生活品质。为此，公园将划定特定区域和设立配套基础设施，确保游客在享受娱乐和教育活动的同时，不损害自然环境。

中国的自然保护区建设历史悠久。早在 1956 年就在广东省建立了全国首个国家级自然保护区——鼎湖山国家级自然保护区。截至 2017 年底，各级政府管理的自然保护区总数已达 2,750 个。据国家林业和草原局的数据，国家级保护区的数量在去年已增至 474 个。

自 2014 年起，中国便开始了国家公园的试点建设，在全国范围内，从青海省到浙江省共设立了 10 个试点公园。国家林业和草原局将国家公园定义为经国家批准并管理的特殊区域，它们通常比自然保护区面积更大，涵盖的生物群落、自然景观以及历史文化遗产（如长城等）也更为广泛。

### 4. English-Chinese Interpreting

**National Parks in the U.S. (The American's Best Idea: Our National Parks)**

Hello, I am from the United States, I actually grew up in the western United States. Today, I would like to talk to you about what has been called American's best idea: our national parks. Ladies and gentlemen, you have probably heard of the Yellowstone National Park. Actually, in the United States, there are more than 400 national parks and monuments today. America was the first country to establish a national park that truly belonged to the people, and that was the Yellowstone. It was officially established as a national park on March 1$^{st}$, 1872. This is when the U.S. Congress passed legislation to create the park, and it was the first of this kind anywhere in the world. I spent a lot of time in the Yellowstone Park as a young boy and had many fun memories in that place. At the north entrance to the park, there is a sign that reads: "For the benefit and enjoyment of the people." This gives you a sense of the importance of national parks in American culture. The establishment of the Yellowstone National Park was the beginning of a movement that actually circled the globe.

Today, more than 100 countries contain some 1,200 national parks and the establishment of Yellowstone National Park was the beginning of a movement, a movement to protect our natural resources, and to protect some of the most beautiful areas not only in the United States, but on the entire planet. That's why many people have said that the national park system of the United States has been American's best idea. Besides the Yellowstone National Park, I have actually had the time and opportunities to travel some of our parks and historic places, places like the Yosemite National Park in California, the Zion National

Park in Utah, the Independence National Historical Park in Philadelphia, Pennsylvania, where the Declaration of Independence was actually signed. Those are just a few of these parks, but there are many more, as we like to say, from coast to coast, or from sea to sea. There are our national parks in Alaska, Florida Keys.

National parks exist for two purposes: one is to preserve the natural wonders of the United States; the other is to make these natural wonders available to all Americans. In the early years of the establishment of national parks, there were advocates who would go and speak with the locals to encourage them to go to these special sites, these new national parks, to they actually believe that if Americans can experience what it is like to go to these parks, for themselves, they would favor the government's support for these parks. So, this is the reason why they persuade people to go and see for themselves.

## 5. Chinese-English Interpreting

普达措国家公园，坐落于滇西北“三江并流”世界自然遗产的核心区域，不仅是香格里拉旅游的璀璨明珠，更是省级自然保护区的重要组成部分。其海拔跨度在 3,500 米～4,159 米之间，以其独特的地理位置和丰富的自然资源，成为了“三江并流”风景名胜区的亮点。

公园内，湖泊湿地如镜，森林草甸葱郁，河谷溪流潺潺，珍稀动植物繁多，原始生态环境得以完好保存。它距香格里拉市城区仅 22 千米，总面积约 1,313 平方千米，是一处集自然风光、生态保护与科学研究于一体的宝地。

普达措国家公园属于亚温带气候，四季分明，秋冬绵长而春夏短暂。夏季凉爽宜人，冬季则稍显寒冷。年平均气温 5.4℃，最热月份（7 月）平均气温 13.2℃，最冷月（1 月）平均气温–3.8℃。在这片土地上，高山湖泊如明镜般映照天空，牧场水草丰美，湿地百花争艳，原始森林中时常可见飞禽走兽的踪迹。

景区内雨量充沛，气候宜人，为植物生长提供了优越条件。这里植被丰富，宛如一座天然的植物园。此外，公园内还分布着多处断层崖、林间小涧、深沟峡谷等独特景观，为地理学研究和旅游观光提供了宝贵资源。

普达措国家公园的植被以冷杉为主，同时还是众多珍禽异兽的家园。其中包括国家一级保护动物黑颈鹤，以及二级保护动物猕猴、猞猁、云豹等。此外，这里还有黑熊、藏鼠兔等经济动物和鹦鹉、画眉等多种观赏性动物，为游客带来无尽的惊喜。

## Related knowledge

### Famous Quotes Concerning Wilderness and the National Parks of the U.S.

In order to assure that an increasing population, accompanied by expanding settlement and growing mechanization, does not occupy and modify all areas within the United States and its possessions, leaving no lands designated for preservation and protection in their natural condition, it is hereby declared to be the policy of the Congress to secure for the American people of present and future generations the benefits of an enduring resource of wilderness.

–Excerpt from The Wilderness Act of 1964

If future generations are to remember us with gratitude rather than contempt, we must leave them a glimpse of the world as it was in the beginning, not just after we got through with it.

–Proclaimed by President Lyndon B. Johnson upon his signing of The Wilderness Act, September 3rd, 1964

There is nothing so American as our national parks... The fundamental idea behind the parks...is that the country belongs to the people, that it is in process of making for the enrichment of the lives of all of us.

–President Franklin D. Roosevelt

The establishment of the National Park Service is justified by considerations of good administration, of the value of natural beauty as a National asset, and of the effectiveness of outdoor life and recreation in the production of good citizenship.

–Theodore Roosevelt

I have always thought of our Service as an institution, more than any other bureau, engaged in a field essentially of morality–the aim of man to rise above himself, and to choose the option of quality rather than material superfluity.

–Freeman Tilden to George B. Hartzog, Jr., ca. 1971

## 口译笔记系统

口译笔记法是口译员不可或缺的核心技能，能帮助口译员精准记录关键信息、记录逻辑与框架和激活短期记忆。口译员需要选择适当的笔记工具记录，借助各种符号系统，如数学符号、偏旁部首、图形符号等，采用纵向、阶梯式的布局方式，记录概念（ideas）、连接词（links）、数字、专有名词等信息内容。

为何要记笔记?

精准记录关键信息：笔记能有效捕捉难以凭借短期记忆保留的数字（如 3,589,024 吨）、专有名词（如人名、地名、国家和组织）以及概念性名词（如“三农问题”“五通一平”“马斯洛需求”等）。

记录逻辑与框架：笔记能清晰地反映信息的逻辑关系和总体框架，通过记录信息间的连接词（如表示框架结构的连接词包括 firstly，first of all，second，secondly 等；表示逻辑关系的连接词 but，however，even though，if，not only...but also 等）来体现。

激活短期记忆：笔记是口译员激活短期记忆的重要工具，帮助他们在翻译过程中迅速回忆起关键信息。

记什么?

笔记内容要点须包括概念（ideas）、连接词（links）、数字、专有名词以及能激活记忆的关键词等。

怎么记?

第一，选择适当的笔记工具：推荐使用带活页圈的上翻式 A6 大小笔记本，便于单手操作；同时，建议选用具有书写流畅且不易漏墨特点的按压式圆珠笔，确保记录效率。第二，采用合适的口译笔记符号。

## 1. 口译笔记符号

口译笔记符号应简洁明了，用最少的符号表达完整意义，代表深层语义而非单个词句。口译笔记符号具有个体性，每位口译员的符号系统可能不同。符号通常对应多个词汇或短语，因为它们代表的是概念而非字面意义。特殊情况下，如专有名词，可以使用缩写表示。口译初学者可参考现成的常用符号，如图形符号、数学符号、偏旁部首、趋向符号等，并根据自身需求和特点进行调整和创新。

## 2. 口译笔记常用符号

（1）图形符号

□ state，kingdom，country，nation，national，federal

△ city，metropolis，cosmopolitans，metropolitan，urban，municipal

▽ village，farm，outskirt，countryside，suburban，rural areas

o 右上角：人

右下角：地方、地名、机构

∧ 领导，监督，顶点，顶级，最，lead，supervise，head，peak，top，supreme，utmost，maximum，climax

△° citizen，urban residents / inhabitants

▽° farmer，peasant，rancher，agriculturist

⊡ universal，worldwide，international，transnational，global，transcontinental，all over the world，across boundaries

→□ 进口 import

□→ 出口 export

（2）数学符号

∵ 因为，由于，幸亏，because，due to，thanks to

∴ 所以，因此，结果，so，therefore，that is why... consequently

+ 和，另外，plus，and，add，more than，besides，moreover，in addition

少+ 更少，fewer

÷ 除了，except

（3）偏旁部首

⺍ 学习，study，learn

宀 家，family，home

忄 快，迅，soon，rapid，swift，speedy，quick，prompt

廴 建设，build，construct

扌 措施，measure

（4）趋向符号（趋向符号通常表示事物的发展趋势）

↑ 增加，上升，完善，繁，increase，heighten，enhance，grow，lift，soar，rise，raise，elevate，ascend，flourish，advance，update，improve，prosperity，boost，give boost to，go up

↓ 减少，降低，贬值，decrease，decline，reduce，descend，diminish，fall，drop，sink，lower，relieve，weaken，lessen，deteriorate，devalue，degrade，depreciate，discount，go down，fall off

↗ 越来，increasing，growing，more and more

大↗ 越来越大，bigger and bigger

⇒ 推动，促进，有助于，promote，strengthen，encourage，stimulate，facilitate，reinforce，strive，push，urge，stir，force，accelerate，drive，prompt，propel，give impetus to，carry forward

⇐ 阻碍，推迟，obstacle，postpone，delay，restrain，withdraw，discourage，curb，hamper，hinder，barrier，bar，encumber，put off，stand in one's way，hold back，draw back

从以上符号中不难发现，一个符号对应了数个词汇或短语，这是因为符号代表的是概念，而不是词句。一些代表专有名词，例如国家、机构的符号除外，通常这些符号用缩写表示。

例如：UN（联合国）—United Nations；UNESCO（联合国教科文组织）—United Nations Educational, Scientific and Cultural Organization；JP（日本）—Japan；USA（美国）—United States of America 等。

### 3. 口译笔记的布局策略

采用纵向、阶梯式的布局方式，这种结构不仅有助于清晰地展现信息间的逻辑关系，还能使整体内容框架一目了然，便于口译员快速回顾和参考。

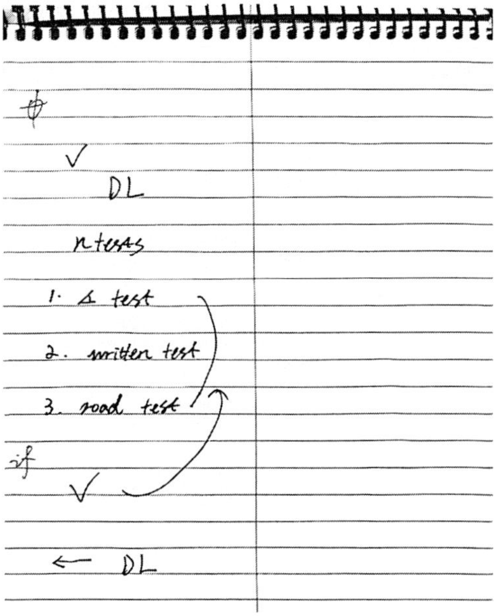

上图笔记原文：To get a driver's license, you must take a series of tests. The first test is the vision test. The second test is a written test. After you pass the written test. The last test you have to take is a road test. If you pass them, you will receive your driver's license.

**1. Sentence Interpreting**

（1）美国的国家公园，每年吸引着数百万游客，它们守护着无数物种的繁衍生息，同时保护着地球上那些令人叹为观止的自然奇观。

（2）国家公园承载着保护自然与文化资源的使命，为当代和后代提供了一个欣赏自然之美、接受教育和启迪的宝贵场所。

（3）从美国犹他州壮丽的红色砂岩到弗吉尼亚州静谧的蓝岭山脉，美国拥有众多丰富多彩、资源富饶的荒野空间。

（4）自 1905 年美国林务局成立以来，政府已历经一个多世纪的时间，致力于公共土地的划定和保护范围的持续扩大。

（5）美国增设数十个国家公园，旨在更好地守护这些珍贵的自然资源，为子孙后代留下宝贵的自然遗产。

（6）National parks refer to the natural areas designated by the state for special protection, management and utilization in order to protect the integrity of one or more typical ecosystems and provide places for ecotourism, scientific research and environmental education. They are distinct from both strict nature reserves and ordinary tourist attractions.

（7）The national park is a type of protected area. It first originated in the United States and has been adopted by most countries and regions in the world.

（8）In 1872, the United States Congress approved the establishment of Yellowstone National Park, the earliest national park in the United States and the world.

（9）The earliest history of the U.S. National Park can be traced back to 1860, and the Yosemite area was designated as the first state park.

（10）In November 2013, the Third Plenary Session of the Eighteenth Central Committee of the Communist Party of China decided to propose the establishment of a national park system for the first time.

**2. English-Chinese Interpreting**

荒野，是那些基本上未受现代人类发展侵扰的广袤土地，通常缺乏道路、建筑等人工痕迹。这些地区为植物和动物提供了自然的栖息地，也是科学家们研究健康生态系统的宝贵场所。

然而，地球上完全原始的荒野已经极为罕见。荒野可以被开发或恢复，其管理方式亦能随时调整。

任何生态系统或大陆，都有荒野地区的存在。例如，澳大利亚大堡礁，作为世界上最大的珊瑚礁，是一片壮丽的水生荒野，庇护着数十种濒危的鸟类、鱼类、哺乳动物和爬行动物。而荒野的范围也各有差异，有的辽阔无垠，如南极大陆的大部分地区；有的则较小，隐匿于意想不到之处，如欧内斯特德布斯区域公园，其位于美国加利福尼亚州人口稠密的洛杉矶市的繁华都市中心。

荒野基金会是一个致力于全球荒野保护的组织，其对荒野的定义包括两个方面：一是某一地方以某种形式的自然生物多样性的存在；二是受到法律的保护，即政府通过立法限制对荒野的开发。

人类与荒野共存了数千年，但随着人口增长和发展加速，许多荒野地区正面临威胁。人类对荒野的影响有直接和间接之分，如营火、偷猎等行为的直接影响，以及气候变化等间接影响，这些都可能限制本地植物的生长或导致非本土物种的入侵。此外，附近的开发活动还可能造成水体和空气污染，威胁荒野生态，甚至阻碍动物迁徙，减少生物多样性。

为了保护荒野，一些政府限制了人们在这些地区的活动，包括采矿、伐木和驾车旅行等。在管理荒野地区时，主要有两种方法：一是保护，通过可持续利用自然资源来维护生态平衡，如限制伐木数量，鼓励补种树苗，或开放部分区域供个人砍伐圣诞树；二是保存，强调通过不利用自然资源来保护荒野的原始状态。

### 3. Chinese-English Interpreting

National parks are representative of the country's most iconic natural and cultural heritage and wildlife habitats. China plans to establish five national parks in the Xizang Autonomous Region, also known as "the third pole of the earth".

The Third Pole National Park will become a national effort to promote eco-tourism on the Qinghai-Xizang Plateau to improve local residents' lives while protecting the environment. The Third Pole National Park will delineate specific areas and infrastructure so that visitors can enjoy entertainment and educational activities without damaging the environment.

In China, there is a long-time history of nature reserve establishment. In 1956, China established the country's first nature reserve–Dinghushan National Nature Reserve in Guangdong Province. By the end of 2017, there were 2,750 nature reserves managed by governments at all levels in China. The number of national-level nature reserves rose to 474 last year, according to the National Forestry and Grassland Administration.

China began establishing with pilot national parks in 2014, and established 10 pilot parks across the country from Qinghai Province to Zhejiang Province. The National Forestry and Grassland Administration defines a national park as one that is approved and managed by the country. Typically, national parks are much larger than nature reserves and cover a wide range of biomes, landscapes as well as historical relics, such as the Great Wall.

## 4. English-Chinese Interpreting

### 美国国家公园

大家好，我来自美国西部，今天我想与大家分享一个被誉为美国人最伟大的创意理念——我们的国家公园。女士们，先生们，或许你们都听说过黄石公园，但美国如今拥有超过 400 个国家公园和纪念地。1872 年 3 月 1 日，黄石公园正式成为世界上第一个真正属于人民的国家公园，这是美国国会通过立法创建的，开创了全球先河。我童年时在黄石公园留下了许多美好的回忆。公园北入口处的牌子写着“为了人民的利益和乐趣”，这充分体现了国家公园在美国文化中的重要地位。黄石公园的建立，标志着一场全球保护自然资源运动的开始。

如今，全球已有 100 多个国家建立了约 1,200 个国家公园。黄石公园不仅是美国自然资源保护的起点，更是全球最美丽地区保护运动的开端。许多人因此认为，美国的国家公园制度是美国人最杰出的创意之一。除了黄石公园，我还有幸游览过美国其他一些令人惊叹的公园和历史名胜，如加利福尼亚的优胜美地（约塞米蒂）国家公园、犹他州的锡安国家公园，以及宾夕法尼亚州费城的独立国家历史公园，那里正是《独立宣言》的签署地。这些只是冰山一角，美国从东海岸到西海岸，从大西洋到太平洋，处处都有美丽的国家公园。甚至在遥远的阿拉斯加和佛罗里达礁岛群，我们也能找到国家公园的踪迹。

国家公园的建立有双重目的：保护美国的自然奇观，并让所有美国人都能亲身体验这些奇观。国家公园的倡导者们曾积极与当地社区沟通，鼓励人们去这些新建立的国家公园。他们深信，只要美国人能亲自去公园感受一番，他们就会支持政府对这些公园的保护。因此，他们强烈推荐大家亲自去体验这些自然奇观。

## 5. Chinese-English Interpreting

Potatso National Park, located in the center of the Three Parallel Rivers World Natural Heritage in Northwest Yunnan, is one of the main tourist attractions in Shangri-La. The park stands between 3,500–4,159 meters in elevation. It is a provincial nature reserve and an important part of the Three Parallel Rivers scenic spot, with its distinctive geographic features and rich natural resources.

Potatso National Park harbors lakes and wetlands, forests and meadows, valleys and streams, rare animals and plants, etc., and its original ecological environment is well preserved. It is 22 kilometers away from the urban area of Shangri-La City, with a total area of about 1,313 square kilometers. It is also a treasured place that integrates natural landscapes, ecological conservation, and scientific research.

The protected area features a sub-temperate climate, with long autumn and winter in contrast to short spring and summer. In summer it is pleasantly cool, while in winter it is a slight chill. The annual mean temperature is 5.4℃, the hottest month (July) has an monthly mean temperature of 13.2℃, and the coldest month (January) has an monthly mean temperature of –3.8℃. In Shangri-La Potatso National Park, there are crystal alpine lakes, pastures with lucid water and grass, wetlands with blooming flowers, and virgin forests where birds and animals often appear.

The scenic spot has abundant rainfall and pleasant climate. Such natural conditions make the plants and vegetation grow luxuriantly, just like a natural arboretum. In addition, there are many unique landscapes such as fault cliffs, forest streams, abysses and canyons, which have extremely high geographical scientific value and tourism appreciation value.

The vegetation in Potatso National Park Reserve is dominated by *Abies fabri*, and there are many rare birds and animals. It is home to black-necked cranes, first-class national protected animals, and second-class protected animals, such as *Macaca mulatta*, lynx, *Neofelis nebulosa* and so on. There are also some economic animals and ornamental animals, mainly *Ursus thibetanus*, *Ochotona thibetana*, parrots and thrushes with high ornamental value, bearing endless delight for the tourists.

# Unit 6 Wildlife Conservation
# 野生生物保护

### What is Wildlife Conservation?

Wildlife conservation is a practice in which people attempt to protect endangered plant and animal species, along with their habitats. The goal is to ensure that nature will be around for future generations to enjoy, and to recognize the importance of wildlife and wilderness lands to humans. Many nations have government agencies dedicated to this practice, and they can help to implement policies designed to protect wildlife. There are also many independent nonprofit organizations that promote various conservation causes.

A number of different disciplines are involved in wildlife conservation. In order to practice it, people must use biology and other sciences to identify populations that are endangered, and to study those populations to learn more about their needs. Economics often becomes involved when organizations work to set land aside for the use of wildlife, with conservationists attempting to arrive at efficient land-use solutions. The field also relies heavily on education, using outreach programs to teach people about wildlife and to show people why conserving natural habitats is important.

Wildlife conservationists work all over the world to identify species that are in need of assistance and to protect them. This discipline often involves capturing animals and breeding them in captivity to ensure that the population remains large and diverse while conservationists work to establish territory for the animals so that they can have a safe place in the wild. Conservation also needs to be balanced with other land uses. Many nations, for example, value forms of recreation in the wilderness, like hiking, camping, and hunting, making it difficult to set aside land specifically for the use of animals.

## Glossary

| | |
|---|---|
| conservationist | （自然环境、野生动植物等）保护主义者 |
| species extinction | 物种灭绝 |
| International Union for Conservation of Nature (IUCN) | 国际自然保护联盟 |
| rhino | 犀牛 |
| domino effect | 多米诺骨牌效应 |
| vegetation zone | 植物带 |
| key protection | 重点保护 |
| amphibian | 两栖动物 |
| reptile | 爬行动物 |
| vegetation type | 植被类型 |
| vegetation restoration | 植被恢复 |
| threatened species | 受威胁物种 |

| | |
|---|---|
| habitat destruction | 生境破坏 |
| bonobo | 倭黑猩猩 |
| community | 群落 |
| biodiversity monitoring system | 生物多样性监测系统 |
| predator | 捕食者、猎手 |
| leopard cat | 豹猫 |
| corsac fox | 沙狐 |
| Yangtze finless porpoise | 长江江豚 |
| hornbill | 犀鸟 |
| stripe-backed weasel | 黄鼠狼 |
| marbled polecat | 虎鼬 |
| species introduction | 引种 |
| The Red List of Threatened Species (The Red list) | 濒危物种红色名录 |
| Ant Forest | 蚂蚁森林 |
| joint patrol | 联合巡护 |
| bank beaver, otter | 水獭 |
| The Endangered Species International | 濒危物种国际组织 |
| ecosystem service | 生态系统服务功能 |

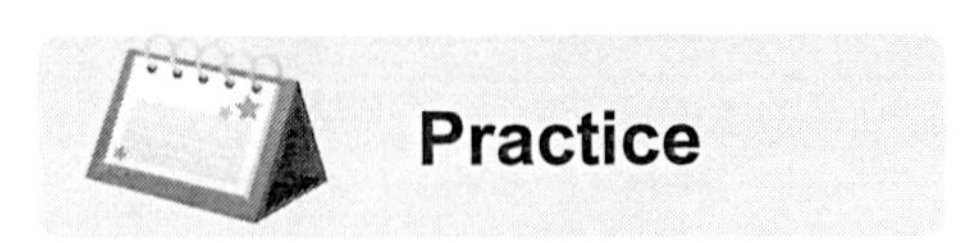

## Practice

### 1. Sentence Interpreting

（1）Africa is home to many endangered or threatened species that include

Western African giraffes, mountain gorillas, bonobos, white rhinos and elephants.

（2）Most of these beautiful endangered wildlife in Africa have been hunted to extinction for their precious body parts, but these special and important animals deserve to live and procreate just like humans. When one of them is killed, it can be very scary for those left behind and difficult for them to survive.

（3）According to the International Union for Conservation of Nature, there are over 19,000 species currently threatened with extinction and the problem is just getting worse.

（4）The IUCN Red List shows an upward trend in the number of endangered animals since it was first released on the World Wide Web in the year 2000. One in four mammals, one in eight birds, one third of amphibians and around 70% of the world's plant life are reportedly under significant threat of extinction.

（5）An invasive species is any non-native plant or animal entering into an environment and unbalancing it in some way. The problem is commonly caused by mistakes in international trade and transport.

（6）深化对野生生物保护的理解，令人惊讶的是，简单的口头传播便能产生显著影响。拯救濒危物种的有效途径之一就是持续不断地谈论它们。一旦你了解到某种动物的生存困境或某种树木的濒危状态，请积极地将这些知识分享给周围的人。无需大规模行动或资金募捐，仅仅通过传播信息，我们就能够为保护野生生物贡献一份力量。

（7）各国政府职能部门有责任确保濒危动物成为学校日常教育的重要话题，同时，各国议会或立法机关应当尽快明确并落实动物保护法。

（8）科学证据已明确指出，杀虫剂的使用与物种丧失之间存在显著关联。在农业生产中，使用除草剂和杀虫剂具有极大的吸引力，许多种植者认为使用这些是有必要的。

（9）据濒危物种国际组织的数据，全球每年消耗约 2.6 亿吨塑料，这些塑料占据了海洋垃圾的 60%至 80%，且部分塑料能在环境中滞留数百年之久。因此，我们的日常垃圾丢弃行为正在不断伤害动物，甚至导致它们误食这些塑料垃圾。

（10）回收利用是保护环境的一项最简便的方式，几乎任何能够降解为单一材料的物品都可以被回收。若更多人参与回收，那么生产产品所需的资源将会大幅减少。

**2. English-Chinese Interpreting**

Nujiang Lisu Autonomous Prefecture of Yunnan Province has abundant and unique biological resources, and the city and three counties of it are all national key ecological function areas. During the 13th Five-Year Plan period, its forest cover exceeded 78%, and the area protected by the Ecological Conservation Redline (ECR) accounted for 61.81% of China's total land area. The populations of endangered animals, such as *Rhinopithecus strykeri* monkey, were increasing steadily, making its protection rate of wildlife under state key protection reach 100%.

Gaoligongshan National Nature Reserve (Northern Section) and Yunling Nature Reserve are all located in Nujiang Prefecture, covering more than 400,000 hectares. There are more than 4,300 species of seed plants, 24 species of the first and second class national protected plants, 580 species of vertebrates and 67 species of animals that under the first and second class national protection in this prefecture. Its species of animals and birds account for more than half of that in Yunnan, and the amphibians and reptiles account for 1/5. Vegetation types, subtypes and community types account for 4/5, 1/2 and 2/5 of the province, respectively.

To further protect the Yunnan snub-nosed monkeys, the prefecture actively carried out 1,550 *mu* of vegetation restoration in the habitat and 810 *mu* of Ant Forest Afforestation Program. The prefecture has conducted joint patrols to speed up the establishment of trans-regional conservation system for Yunnan snub-nosed monkey.

## 3. Chinese-English Interpreting

大家下午好！我很高兴能参加本次“中非野生动植物保护论坛”。

尊重自然，珍爱生命，是中华民族文化的一贯理念和抱负。中国传统文化把自然界和人类看作一个和谐的整体，强调两者之间的和睦共处关系。建设美丽中国，是我们努力奋斗的客观要求和积极追求。它还为我们指明了未来人类命运共同体的方向，朝着一个清洁美丽的世界迈进。

中国政府历来高度重视野生动物保护。在立法方面，中国政府已经通过了多项关于野生动物保护的法律法规。近年来，中国政府实施了生物多样性保护重大工程，加强了对野生动植物进出口的管控。如今，中国政府已经采取了世界上最坚定的立场，在野生动物保护上持零容忍态度。

中国已经与南非、肯尼亚等国家签署了野生动植物保护合作协议，旨在共同加强野生动植物的保护工作。为了支持坦桑尼亚、津巴布韦、博茨瓦纳等非洲国家提升野生动植物保护能力，中国提供了关键的集装箱监控设备以及野生动植物保护设备。此外，中国还积极开展培训项目，旨在加强非洲国家在执法和野生动物保护方面的能力建设。中国政府愿与美国、英国、法国、国际刑警组织等国际组织和非政府组织继续深化合作，共同探讨南非和其他非洲国家在野生动植物保护方面的联合执法机制。我们坚定致力于终结针对大象和其他濒危物种的非法贸易和犯罪行为。

我们非常高兴地看到，通过我们不懈的协作与努力，如今在南非和非洲地区，人们听到的关于中国人偷猎大象和犀牛，或走私象牙和犀牛角制品的案例已经显著减少。然而，尽管我们取得了显著进展，但我们也要意识到，仍然有可能存在一些中国人在利益的驱使下从事野生动植物及其产品走私及非法盗猎的犯罪活动。这些行为严重损害了中国和中国人民的国际形象。我想向大家明确保证，中国政府和中国人民对野生动物保护的坚定承诺是决不动摇和妥协的。我们重申，在涉及中国公民时，无论是在中国境内还是在其他任何地方，中国政府都将秉持“零容忍”的态度，坚决反对并严厉打击任何针对野生动植物的犯罪行为。

贫穷是世界公敌。同时，贫穷也是野生动物保护的头号敌人。令人悲哀的是，在非洲的某些地区和部分国家，有很多无业人员被迫盗猎野生动物来谋生。的确，加强保护意识和执法很重要。但是，对那些人而言，更重要的是，要通过创造就业岗位，使他们能够从事正常的经济活动。因此，我们必须采取短期和永久的方案，来解决野生动物保护难题。

我们愿与南非的朋友们携手合作，紧抓这一重要的历史机遇，深度融合我们各自的发展战略，进一步深化跨越国境的全面互利合作，并为这一伙伴关系注入新的活力。我们坚信，通过共同努力，我们不仅能够推动中南全面战略伙伴关系迈向新的高度，还能够为中国与南非两国人民带来更多的福祉，为野生动物保护做出新的贡献。

最后，预祝本次大会圆满成功！谢谢大家！

### 4. English-Chinese Interpreting

The amended national list on protected wild animals was announced. There are 988 species of wild animals on the new list, 517 more than the previous version that was released over 30 years ago. The new list represents major progress as it views animals as a part of a healthy ecological system, instead of their impact on human society.

For example, wolves have been added to the national second–class wildlife protection list. For a long time, wolves have been viewed as “vermin” as they came into conflict with human herders. Yet now it is a wide consensus among both scientists and the public that predators such as wolves are also needed for a healthy ecological system. Besides large predators, smaller ones such as leopard cats, corsac foxes and racoon dogs have all been added to the list. For decades, predators have been the group of animals most affected by shrinking habitats because of human activities. By putting them on the protection list, the move will help better protect them and thus the whole ecological system.

Several species that have gained wide public attention, such as finless porpoises and several kinds of hornbills, have all been added to the list or promoted from the second to first-class state protection list. That is doubtlessly a reward for the dedication of those researchers who have been calling for better wildlife protection for years.

Of course, there is much to improve. For example, stripe-backed weasels and marbled polecats are even rarer than the listed animals but there is little research about them, and no one even knows how many of them still exist. More research is needed to call for more attention to them. And other animals such as turtles and bank beavers also need more attention, as they are of tremendous importance to the protection of river ecosystems. It is hoped that more research can be done on them, so they too can be added to the list.

## 5. Chinese-English Interpreting

根据《地球生命力报告 2018》的数据，对全球 4,005 种脊椎动物中的 16,704 个种群进行的生物多样性丰富度监测表明，2014 年的地球生命力指数与 1970 年相比，已显著下降了 60%。与《地球生命力报告 2016》的数据相对比，2012 年的地球生命力指数相较 1970 年已下降了 58%。特别值得警惕的是，由于人类的过度消耗，已有 300 种哺乳动物面临灭绝的威胁，特别是海洋生物因过度捕捞而遭受巨大压力。这些触目惊心的数字不仅反映了地球生命力指数的显著下降，更警示我们这种下降趋势仍在继续。

报告强调，一旦生物多样性遭受破坏，其所支撑的关键生态系统服务功能，如粮食生产、纤维供应、医药资源、淡水供给、作物授粉、污染物过滤以及自然灾害的防御等，都将面临严重威胁。若生物多样性丧失，人类将面临严重的食物短缺问题，清洁水源将变得稀缺，空气质量将不断下降，而世界将被各种废弃物所充斥。

《地球生命力报告 2018》明确指出，自公元 1500 年以来，全球范围内所有灭绝的植物、两栖动物、爬行动物、鸟类和哺乳动物中，高达 75%的物种灭绝归因于人类的过度开发、农业活动，或这两者的共同作用。

人类过度索取自然资源，不断挤压野生动物的生存空间，最终这种行为也“反噬”了人类自身。从 SARS、MERS、新冠病毒、埃博拉、禽流感到猪流感，这些疾病的根源无一不指向野生动物。人类的不当行为加速了野生动物的急剧减少和物种灭绝，同时也为寄生于这些动物身上的病毒提供了跨越物种传播的机会，使病毒进入人类世界。

从更深层次审视人与野生动物的关系，物种灭绝的四大主要原因可归结为：过度猎杀、生境破坏、盲目引种和环境污染。在这些因素中，人类活动以及对自然的过度开发对物种构成了最直接且严重的威胁，特别是栖息地的丧失和退化。

## Related knowledge

### Why Can't We Clone Endangered Species to Save Them?

Tigers, gorillas, rhinos, it seems like there are a lot of animals on the endangered species list these days. But we know how to clone animals, so, if the problem is too few rhinos, and if we know how to make more rhinos, problem solved, right? Well, there are actually people working on cloning endangered animals. However, it is not as easy as you might think, and in the end, it still won't fix the problems that made those species endangered in the first place.

The first mammal cloned from adult cells was Dolly the sheep back in 1996, and since then, different animals like cats, dogs, and even monkeys. And it didn't take long for endangered animals to get on that list. In the early 2000s, for example, researchers cloned a type of wild cow called the gaur, and they went on to clone other endangered cattle, wild sheep, and even an extinct wild goat. However, as impressive as these achievements were, none of them ended up really increasing the species' numbers.

That's because, despite what movies might suggest, you can't just order up an army of clones. To clone something, you typically need more than just the DNA from the original creature. You also need an egg cell to put that DNA into to create an embryo. Then that embryo needs to be implanted in a surrogate mother to develop. You can't just do this in a petri dish. While that all might sound straightforward enough, getting those eggs in the first place can be difficult and risky for the animal. To know how to get eggs, you need to know a lot about that specific animal's reproductive biology, like when they ovulate, or grow and release egg cells.

Humans generally ovulate in regular, frequent cycles, but not all animals do. For example, pandas only ovulate once a year, and there are animals like cats that only ovulate after mating. We can sometimes use hormone injections to kick start ovulation, but each animal may need different doses or a different mix of hormones. And then you still have to get the eggs. Take rhinos, for example, you try convincing a 1,700 kilogram ovulating rhino to let you collect eggs from inside her using a custom-designed, one-and-a-half-meter long device. I won't. I mean sure, you can sedate her. Which is definitely what the researchers do. But sedation also carries risks. All and all, getting eggs from animals may involve multiple hormonal injections and the use of anesthetics or surgical procedures, and not every individual animal is healthy enough to undergo all that. And also, you need to collect a lot of eggs. The DNA-egg cell fusion doesn't always work right, so for each viable embryo, you might need hundreds of egg cells.

Even if you go through all of that, you still need surrogates to carry the pregnancies. Like collecting eggs, that requires knowledge about the animal's reproductive biology, as well as lots of healthy female animals; more than just the number of clones you want to make. Unfortunately, not every implantation takes and develops into a healthy offspring. For plentiful species, or ones we know really well, like cows or sheep, those might not be big problems. But for endangered species, like tigers or rhinos, there may not be enough female animals available to process. The good news is that there are ways to kind of cheat. You can sometimes use eggs or mothers from more common, closely related species, for example. The baby gaur's surrogate mom was actually a regular cow named Bessie. Because, of course, all cows are named Bessie, apparently. But hybrid embryos can have extra problems during development. Scientists are exploring on an alternative to all this: a way to make eggs and sperm by genetically reprogramming frozen tissue samples. So one day we might be able to make a lot of eggs without needing a lot of female animals. But that's still in the very early stages, and the process would likely need to be customized for each animal.

Now, it's possible that we might figure out how to solve all of these challenges. Nevertheless, even if we could clone any animal we wanted, it might not be the solution we were looking for. For one thing, even though cloning might increase the number of animals, you could still end up with a loss of genetic diversity because they'd all be genetically identical. That's bad because the population could effectively become inbred, which can leave them vulnerable to diseases or genetic disorders. If you have frozen tissues from a lot of different individuals, you might be able to avoid this issue or even reintroduce genetic variation that's been lost, but that will depend on what you've stockpiled. The real problem with cloning endangered animals is that it won't stop the poachers, habitat loss, or myriad other things driving these species extinct in the first place. Cloning is very expensive, and some researchers have pointed out it's probably cheaper and more effective to spend that money on fighting poaching, creating new nature reserves, or other, more traditional conservation efforts. Still, scientists working on cloning remain optimistic that it could one day become a really useful tool for conservation, like to reintroduce lost genes or a last lifeline to save species whose numbers have dwindled to just a handful.

# 口译应对策略

口译是一种多任务同时进行的活动（multitasking），口译工作中难免遇到听不懂、听不清、记不住、说不出等情况，这些是每一个口译员都会遇到的问题。遇到这些问题要镇定，积极寻找解决方法。口译应对策略包括数字模糊处理、原语复读、信息省略、解释、请人帮助、合理询问和口译失误处理。

## 1. 数字模糊处理

对于数字的口译，只要超过万或万亿以上的计量级别，可进行模糊处理。有时口译员会面临这样的困境：既要处理复杂的句子逻辑和语言结构，又要再现精确的数字，应接不暇之际又必须以极快的速度对源源不断的信息进行产出。在这种情况下，只得把听见的数字以近乎四舍五入的方法处理成大约数，如“63.2 万”处理成“over 600,000”，“7,630 亿”处理成“760 billion”等。这是大脑寻求自我保护的一种解脱方法。有时候，如果在翻译时真的给出非常准确的数字，反而会让外国听众一头雾水，他们会对你花了九牛二虎之力翻出来的数字报以“so what?”的感慨，因为通常在这些具体数字之后，发言人会对这些数字进行分析，如 A 数字比 B 数字增长或减少多少百分比，这才是听众关心的。数字的绝对价值远没有数字的相对比较价值有意义，在会议口译场合中尤其如此。翻译数字需要抓大放小：虽然不是很精确，但的确能够最大限度保留原文的意思，比费劲正确翻译出一个具体数字但遗漏接踵而来的数字或信息的“性价比”更高。

## 2. 原语复读

对于某些术语、专有名词、人名等词汇，由于时间有限，不知道对应的译文表达，或由于根本不理解原语意义以至于无法给出译文时，可采用原语复读的方法。

例如：UFO，武术，旗袍，阴阳，太极，amino acid，hydrogen，chloride，coelacanth，Mr. Ezra，Aruba（地名），Brunei Darussalam（地名）和 BUS（数据总线）等。

### 3. 信息省略

口译内容主要是提取和归纳源语重点信息，以下情况可以省略不译：

（1）发言人太啰唆，不断重复。

（2）听众无法理解或者会误解的源语信息。例如中国文化讲究谦虚，这种谦虚会体现在演讲中，例如一段中文开场白："谢谢大家光临，由于本人学识浅薄，加之准备不充分，所讲之处肯定有很多疏漏和不妥，请大家包涵。"译员不必把所有信息都翻译出来，只用翻译出最主要的信息"Thank you for coming"。后面带有文化色彩的信息如果全部都翻译出来，西方听众会觉得很困惑，甚至误以为演讲者无能，无资格进行发言。

### 4. 解释

当译员对某一表达无法找到确切的译语表述时，可通过解释的办法将信息传达给听者。

例如：

三农问题：Issue related to agriculture, rural areas and farmers

夫妻肺片：Sliced Beef and Ox Tongue

小葱拌豆腐，一清二白：explicit/well-understood

Armed to teeth：武装到牙齿，全副武装

Achilles's heel：致命弱点

As gay as a lark：兴高采烈

解释需要建立在理解原文的基础上，如果不理解则无法解释说明，千万不能根据字面意思编造，应立刻向说话人询问原文的意义。

## 5. 请人帮助

交替传译通常只需要一名口译员进行现场翻译，没有同伴。如果在翻译中遇到无法理解或不认识的术语，可以当场请教发言人或台下的观众。台下的观众大多是相关领域的专家，在这方面的语言知识说不定比口译员更丰富。

## 6. 合理询问

由于某些环境因素，口译员无法与观众直接面对面交流，而又遇到自己无法理解或听不懂的词句，这时应合理询问发言人，不能抱有侥幸心理胡乱猜测。以下是一些提问方法：

Sorry, sir?

Sorry, I missed that. Could you say that again please?

Sorry, I didn't catch that. Could you repeat it, please?

Sorry, I don't quite see what you mean. Could you just repeat that, please?

有时，如果发言者说了一段较长的话，而口译员只是没听懂其中一部分信息，因此光说自己没听懂还不够，还需具体说明是哪个地方没听懂，例如：

There's a good chance that … Sorry, the good chance is what?

The whole project needs … Sorry, the whole project needs what?

The first phase, from 1978 to 1984, concentrated on … Sorry, concentrate on what?

如果发言过程中无法打断发言人，可等到讲完后再问。此时，要先界定问题的范围再提问，例如：

Focus: Sorry, sir, you talked about Britain hopes to achieve the target saving…

Clarification: Could you repeat that part please?

Context: Excuse me, sir, when you were talking about economic crisis…

Focus: You mentioned about China…

Clarification: Could you clarify your point/position on that?

以上是合理询问的主要方式。注意：①如果问题太多，会占用会议时间，听众或客户会对口译员失去信任，认为该口译员并不专业。因此，在提问之前，要问自己，这个问题是否是重要信息？是否影响对发言的后续理解？是否真有必要提问？问完之后，一定要集中注意力听取，千万不要问了一遍没听懂，再问一遍，这样会使口译员形象大打折扣。②提问要礼貌，并记得道谢。

**7. 口译失误处理**

口译工作中的失误包括三种情况：口译员翻译失误、发言人失误和口译设备失误。

- 口译员翻译失误

由于口译的即时性和现场性特点，口译员难免出现错译和漏译的现象。切记不要慌张，个别发音失误可以忽略，只要不是在十分正式的场合，非重要信息误译和漏译也可忽略，只要不影响主要信息的传达，不影响大局。因为口译的主要目的是传递信息而不是按照笔译的要求字斟句酌。如果口译员发现了自己的内容出现了较大失误，影响沟通交流，则需要及时修正。

Source language: Sumatra is the second largest island of Indonesia.

Target language: 苏门答腊是印度尼西亚第一大岛，更正：苏门答腊是印度尼西亚第二大岛。

Source language：东盟是一个极度活跃的区域性组织。

Target language：ASESAN is an extremely active international organization.

Correction: It's a regional organization.

- 发言人失误

演讲具有现场性和即时性特点，发言人出现失误是无可避免的。口译员不应慌张，应冷静处理。主要方法有：自动更正、暗示和询问。对于发言人的口误，以及能导致误解的文化负荷词，口译员应自动更正。如果发言人发现自己的失误并修正，口译员只需译出修正的词句。如果口译员对发言人的失误并无把握，则可在翻译时减缓语速，用疑问的眼神提醒发言人，或者请发言人重复一遍。如果发言人仍维持原话，口译员则可照译。

- 口译设备失误

交替传译需要使用麦克风。口译员在会前要提前到场确保口译员坐席配备麦克风并测试麦克风是否正常。在会议问答环节，如果提问者没有配备麦克风或话筒，口译员应及时提醒。如果没有多余的设备，或接收效果很差，口译员应尽量译出大意，并向听众或主办方说明情况。

## 参考答案

### 1. Sentence Interpreting

（1）非洲是许多濒危或受威胁物种的家园，其中包括西非长颈鹿、山地大猩猩、倭黑猩猩、白犀牛和非洲象。

（2）非洲这些美丽的濒危野生动物往往因它们珍贵的身体器官而遭受猎杀。然而，这些独特且重要的动物与我们人类一样，都应当享有生存和繁衍后代的权利。每当一个物种的个体被猎杀，对于剩下的成员来说，它们所面临的生存环境将变得更加恶劣，生存前景堪忧。

（3）国际自然保护联盟警示，当前已有超过 19,000 个物种处于濒临灭绝状态，且这一形势预计将持续恶化。

（4）《IUCN 红色名录》显示，自 2000 年首次在万维网上发布以来，濒危动物和植物的数量均呈上升趋势。据报道，全球范围内，四分之一的哺乳动物、八分之一的鸟类以及三分之一的两栖爬行动物均面临灭绝的重大威胁。更令人担忧的是，世界上大约 70%的植物也面临同样的命运。

（5）入侵物种是指进入新的环境并以某种方式破坏生态平衡的任何非本地植物或动物。这个问题通常由国际贸易和运输中的失误导致。

（6）It’s amazing how much word of mouth can do to raise awareness about wildlife conservation, and one of the best ways to help endangered species is to keep talking about them. If you know the plight of an animal or the worrisome condition of a tree, pass your wisdom on to others. It doesn’t take a lot, and you don’t need to raise money, just need to publicize.

（7）National government authorities also have a responsibility to ensure that endangered animals are a regular topic of discussion in schools, and national parliaments or legislatures should clarify and implement animal protection laws as soon as possible.

（8）Some scientific evidence suggests a clear relationship between pesticides and species loss. The temptation to use herbicides and pesticides in agricultural production is so great that many growers consider them necessary.

（9）According to Endangered Species International Organization, we use about 260 million tons of plastic every year. Plastics make up 60%–80% of marine litter, and some remained in the environment for hundreds of years. Animals injure themselves every day with our wastes, or eat them unknowingly.

（10）Recycling is one of the easiest things to do to help the environment. Almost anything that can be decomposed into a single material can be recycled. If more people recycle, fewer resources are needed to produce products.

## 2. English-Chinese Interpreting

云南省怒江傈僳族自治州拥有丰富而独特的生物资源，市及其三县均为国家重点生态功能区。“十三五”期间，其森林覆盖率超过 78%，生态保护红线范围占其总面积的 61.81%。怒江金丝猴等濒危动物种群数量逐渐增加，在这里，国家重点保护野生动物保护率达到了100%。

高黎贡山国家级自然保护区和云岭自然保护区均坐落于云南省怒江傈僳族自治州境内，总面积超过 40 万公顷。怒江州内拥有种子植物达 4,300 多种，其中包括 24 种国家一、二级重点保护野生植物，充分展现了其丰富的植物多样性。在动物资源方面，怒江州拥有 580 种脊椎动物和 67 种国家一、二级重点保护野生动物，动物和鸟类的种类占到了云南省的一半以上，两栖爬行动物种类更是占据了云南省的五分之一。此外，怒江州的植被类型、亚型和群落类型也非常丰富，分别占据了云南省的 4/5、1/2 和 2/5 的比例。

为进一步保护滇金丝猴，针对金丝猴的栖息地，该州成功实施了1,550 亩植被恢复项目，以及 810 亩的蚂蚁造林工程。同时还开展了联合巡护，加快构建滇金丝猴跨区域保护体系。

### 3. Chinese-English Interpreting

Good afternoon, everyone! I am very glad to be here to participate in this China-Africa Wildlife Conservation Forum.

Respecting nature and cherishing life is the consistent philosophy and aspiration of the Chinese culture. Traditional Chinese culture regards nature and human beings as a harmonious whole, emphasizing the harmonious coexistence relationship between the two. Building a beautiful China is the objective requirement and pursuit of our hard work. It also gives us the direction of a community of shared future for mankind, moving towards a clean and beautiful world.

The Chinese government has always attached great importance to wildlife protection. In terms of legislation, the Chinese government has passed a number of laws and regulations on wildlife protection. In recent years, the Chinese government has implemented major biodiversity conservation projects and strengthened control over the import and export of wild animals and plants. Today, the Chinese government has taken the toughest stance in the world, with a zero-tolerance approach to wildlife protection.

China has signed cooperation agreements on wildlife protection with South Africa and Kenya. We have provided container monitoring equipment for important wildlife species, as well as wildlife protection equipment to Tanzania, Zimbabwe, Botswana and other countries. China is also actively carrying out training programs to support African countries in strengthening capacity building in law enforcement and wildlife protection. The Chinese government is willing to continue to cooperate closely with the United States, the United Kingdom, France, Interpol and other international organizations and non-governmental organizations to jointly explore the joint law enforcement of wildlife protection in South Africa and other African countries. We will do our best to end crime against elephants and other endangered species.

We are very pleased to see that today in South Africa and Africa we are hearing fewer cases of Chinese poaching elephants and rhinos, or smuggling ivory and rhino horn products, thanks to our concerted and tireless efforts. However, as we continue to move forward, it cannot be completely ruled out that some Chinese people will still smuggle wild animals and plants and their products and carry out illegal poaching crimes driven by profit. Their crimes have seriously damaged the image of China and the Chinese people. However, what I want to assure you is that the firm commitment of the Chinese government and Chinese people to wildlife protection will never waver or compromise. I would like to reiterate that the Chinese government has a "zero tolerance" attitude toward relevant crime involving Chinese citizens, whether in China or anywhere else, and firmly opposes any crime against wildlife.

Poverty is the enemy of the world. Meanwhile, poverty is the number one enemy of wildlife protection. We are very sad to see that in some regions and countries in Africa, many unemployed people are forced to engage in wild animal poaching to make a living. Of course, awareness-raising and law enforcement are important; but, for those people, it is even more important to enable them to engage in normal economic activities through job creation. We believe that we must seek both long-term and permanent solutions of wildlife protection.

We are willing to work together with South African friends to seize important historic opportunities to fully coordinate our development strategy, deepen comprehensive and mutually beneficial cooperation across borders, and inject new vitality to further promote the development of China-South Africa comprehensive strategic partnership. We are confident that our two sides will definitely bring more benefits to the people of China and South Africa, and make new contributions to wildlife protection.

Finally, I wish this conference a complete success! Thank you!

**4. English-Chinese Interpreting**

修订后的《国家重点保护野生动物名录》已公布。新名录共有 988 种野生动物，比 30 多年前发布的上一版增加了 517 种。新名录代表了重大进展，因为它将动物视为健康生态系统不可或缺的一部分，而不再仅仅关注野生动物对人类社会的影响。

以狼为例，它现已被正式列入国家二级保护野生动物名录。长期以来，由于与人类牧民的冲突，狼常被视为“害兽”。然而，如今科学研究人员和公众都普遍认识到健康的生态系统同样需要狼这类食肉动物的参与。除了狼这样的大型食肉动物，豹猫、沙狐和浣熊等较小的食肉动物也已被纳入新公布的国家重点保护野生动物名录。几十年来，受人类活动的影响，食肉动物是受栖息地萎缩影响最严重的动物群体。将其列入保护名录，有助于更好地保护它们，进而保护整个生态系统。

江豚和几种犀鸟等备受瞩目的物种，现已成功被列入国家保护名录，甚至部分物种由国家二级保护名录提升至一级保护名录。这一举措无疑是对多年来致力于野生动物保护研究的科研人员的辛勤付出和不懈努力的肯定与回报，多年以来他们一直坚定地呼吁社会大众加强对野生动物的保护。

诚然，野生动物保护领域仍有诸多待改进之处。以条纹背鼬和虎鼬为例，这两种稀有动物相较于前述物种更为稀有，受到的关注度与研究却严重不足，甚至它们的种群数量仍是未知数。因此，我们迫切需要加大研究力度，呼吁公众提高对他们的关注度。同样，海龟和河狸等动物对河流生态系统的平衡起着至关重要的作用，但它们的保护状况同样不容忽视。我们期待能有更多的研究聚焦于这些动物，并将它们纳入保护名录。

### 5. Chinese-English Interpreting

According to the Living Planet Report 2018, based on monitoring of the biodiversity abundance of 16,704 populations of 4,005 species of vertebrates around the world, the Living Planet Index in 2014 was 60% lower than that in 1970. Compared with the data in the “Living Planet Report 2016”, the Living Planet Index (LPI) in 2012 was 58% lower than that in 1970. Of particular concern is the fact that 300 species of mammals are endangered due to human consumption, particularly overfished marine life. These figures imply a remarkable decline in the Living Planet Index, and more worryingly, it’s a continuing decline.

The report said，ecosystem services such as providing food, fiber, medicine, fresh water, crop pollination, filtering pollutants, and protecting against natural disasters will all be threatened if biodiversity is destroyed. Without biodiversity, human beings will face problems such as insufficient food, lack of clean water, degraded air quality, and the world will be full of various wastes.

The Living Planet Report 2018 specifically mentioned that 75% of all plants, amphibians, reptiles, birds and mammals that have become extinct since AD 1,500 were caused by human over exploitation, agricultural activities or a combination of both.

Human beings' excessive demands on the natural world and intensifying encroachment into the living space of wild animals eventually backfired themselves. SARS, MERS, COVID-19, Ebola, bird flu, swine flu, are all without exception. Human misconduct has accelerated the drastic reduction of wild animals and even the extinction of species, and in the process, it has also increased the possibilities for viruses parasitic on wild animals to spread across species, bringing the virus to the human world.

Looking at the relationship between humans and wild animals from a deeper level, there are four major reasons for species extinction, namely overharvesting, habitat destruction, invasive species, and pollution. Human activities and their over exploitation of nature directly constitute the greatest threats to species, including habitat loss and degradation.

# Unit 7 Horticulture
# 园　艺

## Introduction

Horticulture is an unfamiliar term to many people. Yet, when they are told that caring for houseplants, growing tomatoes in the garden, and mowing the lawn are horticultural activities, the term suddenly becomes much clearer. Give it some thought, and you'll find that horticulture is part of our culture and everyday lives. Horticulture is a segment of the agriculture industry. The term horticulture literally means the culture of a garden. However, the term has taken on a broader context. Horticulture includes the production and use of plants for food, comfort, and beautification. A direct relationship exists between horticulture and science. The area of science most closely associated with horticulture is botany. Botany is the study of plants and plant processes. The field of science that deals with the cultivation of horticultural plants is known as horticulture science. Science is applied across the horticulture industry. The application of science to horticulture is called horticulture technology. Successfully raising horticultural plants takes more than just daily watering. Time, patience, and an understanding of diverse scientific processes are needed to produce optimal plant growth.

The horticulture industry is the combination of scientific, technological, and production activities that ensure the satisfaction of the consumer. The horticulture industry can be divided into three areas: pomology, olericulture, and ornamental horticulture. Each area is unique and includes many career opportunities.

The growth and use of plants for their beauty is the area of horticulture known as ornamental horticulture. Ornamental horticulture involves the production and use of woody and herbaceous plants. Because of the use of green plants, the ornamental horticulture industry is often referred to as the "green industry." Ornamental horticulture is divided into two categories. They are floriculture and landscape horticulture. Both involve the use of flowering and foliage plants. Foliage plants are those used for their colorful greenery or leaves.

The production and use of plants to beautify the outdoor environment is landscape horticulture. Landscape horticulture includes designing plans for landscapes, installing landscapes as specified in the plans, and maintaining the landscapes. Maples, oaks, forsythia, honeysuckle, English ivy, peonies, and Kentucky bluegrass are common landscape plant materials.

Most plants used in landscape horticulture are obtained from a nursery. A nursery is a place that specializes in starting plants and growing them until they are ready to be transplanted to landscapes. Nurseries produce ground-cover plants, herbaceous perennials, flowering shrubs, evergreens, deciduous shade trees, and ornamental trees for retail and wholesale purposes. Other nurseries, sometimes called sod farms, grow grass that is cut and moved to landscape projects.

## Glossary

| | |
|---|---|
| horticulture | 园艺（学） |
| International Union for the Protection of New Varieties of Plants (UPOV) | 国际植物新品种保护联盟 |
| fresh cut flower | 鲜切花 |
| plant breeding | 植物培育 |
| plant breeder | 植物育种者 |
| pomology | 果树栽培（学） |
| olericulture | 蔬菜栽培（学） |
| floriculture | 花卉栽培（学） |
| maple tree | 枫木 |
| forsythia | 连翘属植物 |
| honeysuckle | 金银花 |
| English ivy | 英国常春藤 |
| peony | 牡丹 |
| Kentucky bluegrass | 肯塔基蓝草 |
| nursery | 苗圃 |
| ground-cover plant | 地被植物 |
| herbaceous perennial | 多年生草本植物 |
| flowering shrub | 开花灌木 |
| evergreen | 常青树 |
| deciduous shade tree | 落叶乔木 |
| gerbera | 非洲菊 |

| | |
|---|---|
| gypsophila | 满天星 |
| lisianthus | 洋桔梗 |
| carnation | 康乃馨 |
| hydrangea | 绣球 |
| variety | 品种 |
| potted plant | 盆栽植物 |
| plant variety protection (PVP) | 植物品种保护 |
| stem | 花枝、枝干 |
| soilless cultivation | 无土栽培 |
| gardener | 园丁、园艺师 |
| Aalsmeer | 阿尔斯梅尔（荷兰西部城市） |
| cultivar | 培育品种 |

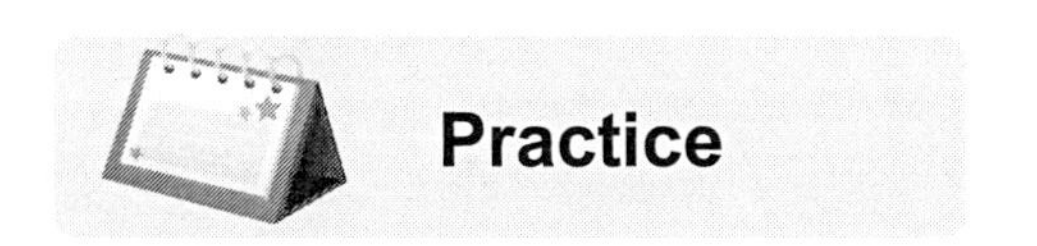

## Practice

### 1. Sentence Interpreting

（1）The International Union for the Protection of New Varieties of Plants (UPOV) is an intergovernmental organization with headquarters in Geneva, Switzerland.

（2）UPOV was established by the International Convention for the Protection of New Varieties of Plants. The Convention was adopted in Paris in 1961 and it was revised in 1972, 1978 and 1991.

（3）UPOV's mission is to provide and promote an effective system of plant variety protection, with the aim of encouraging the development of new varieties of plants, for the benefit of society.

（4）The term “species” is a familiar unit of botanical classification within the plant kingdom. However, it is clear that within a species there can be a wide range of different cultivars. Farmers and growers need plants with particular characteristics and that are adapted to their environment and their cultivation practices.

（5）A plant variety represents a more precisely defined group of plants, selected from within a species, with a common set of characteristics.

（6）经过三十年的蓬勃发展，云南花卉产业在产业规模、市场供给、品种多样性及市场覆盖上均领跑全国。

（7）近年来，花卉消费的不断升级推动了鲜花消费的家庭化和日常化，预示着中国花卉市场巨大的消费潜力将持续激发云南花卉产业的迅猛增长。

（8）随着 5G 时代的到来，花卉产业的数字化转型升级已成为刻不容缓的发展趋势。

（9）受益于进出口免税等优惠政策以及国家产业集群、省级“一县一业”示范创建等重大项目的推动，云南花卉产业正积极探索一条高质量、高效率的发展新路径。

（10）云南花卉产业在新优品种的引进、育种和推广，以及无土栽培、市场营销、品牌推广等方面均取得了显著成效。

## 2. Chinese-English Interpreting

### 第 21 届中国昆明国际花卉展致辞

尊敬的各位嘉宾，女士们，先生们：

大家上午好！

非常高兴出席第 21 届中国昆明国际花卉展。在此，我谨代表中国花卉协会，对国际花卉展的隆重举行表示热烈的祝贺！向长期以来关心、支持我国花卉事业发展的各级领导、社会各界朋友表示衷心的感谢！

昆明自然条件优越，花卉业发展久负盛名，已成为享誉全球的花卉之都。中国昆明国际花卉展已成功举办 20 届，是我国花卉界具有重大影响力的花事活动，在展示新成果、促进交流合作、推动花卉贸易等方面发挥了积极作用。

云南省委、省政府高度重视花卉产业发展，花卉种植面积和销售额长期位于全国前列，探索出了许多好的经验和做法，在全国发挥了示范引领作用。

同志们！我国花卉植物资源丰富，花卉栽培历史悠久，花文化底蕴深厚。经过多年的努力，我国花卉业得到了长足发展，已成为全球最大的花卉生产国、重要的花卉消费国。但与花卉业发达国家相比，与美丽中国建设和人民群众对美好生活的需求相比，还有很大差距。

去年11 月和今年 5 月，国家林业和草原局、农业农村部明确要求：要“以推进花卉业高质量发展为目标”“着力构建战略布局合理、产业链条完整、创新迭代活跃、质量高效益好、区域协调发展的现代花卉产业体系”，提出了“到 2035 年基本建成花卉强国”的发展目标。花卉行业必须要在“高质量发展”“现代花卉产业体系”“建成花卉强国”这三个方面系统谋划，积极推进。

构建现代化花卉产业体系是实现花卉业高质量发展、建设花卉强国的必然要求，是实现花卉产业现代化的重要支撑，是新时代赋予花卉行业的新使命。面对新形势、新要求，花卉行业必须要加大花卉种质资源保护力度，建立健全保护体系和保护机制，确保我国花卉种质资源安全；必须要强化花卉种业自主创新，建立花卉种业创新体系，提升我国花卉品种自主化水平；必须要肩负起新时代赋予花卉行业的历史使命，更好地服务生态文明和美丽中国建设。

云南省自然条件优越，花卉资源丰富，产业发展基础坚实。在推进花卉业高质量发展的新形势下，我们殷切地期待云南省在推进现代化花卉产业体系建设中走在前、做示范，创造更加辉煌的业绩。

最后，祝本届花卉展圆满成功！

谢谢大家！

### 3. English-Chinese Interpreting

It is humbling to consider the wonder of plants and to appreciate their structure and function combining engineering and chemical manufacture beyond the dreams of man: highly efficient, fully coordinated and multi-functional organisms with no moving parts and fuelled by natural resources. It is said that environmental factors drive evolutionary responses and this is clearly demonstrated by the infinite variations found in plants to exploit environmental niches. Gardeners need to appreciate that this adaptation to environment is not for decoration or their delectation but for the plant's function and survival.

Understanding the science underpinning the differences between form, function and survival strategies of different plant groups will greatly aid their cultivation and the enjoyment of gardening. The relationship between plants and the environment in which they grow is dynamic. The plant can obviously respond immediately to short-term stress caused by factors such as variation in water supply and temperature, but it also must often be sensitive to regular seasonal changes to trigger major physiological processes such as the change from vegetative growth to flowering. Much is already known about these processes and is incorporated into horticultural practice, but the further understanding of these mechanisms, and of the trigger signals that initiate them, is an exciting area of science that will have great implications for the gardener and the commercial grower.

Horticulture includes the production and use of plants for food, comfort, and beautification. The area of science most closely associated with horticulture is botany, which is the study of plants and plant processes. The field of science that deals with the cultivation of horticultural plants is known as the horticulture science. The application of science to horticulture is called the horticulture technology. The horticulture industry can be divided into three areas: pomology, olericulture, and ornamental horticulture. Pomology is the planting, harvesting, storing, processing, and marketing of fruit and nut crops. Olericulture includes the planting, harvesting, storing, processing, and marketing of vegetable crops. The growth and use of plants for their beauty is the area of horticulture known as ornamental horticulture. Ornamental horticulture is divided into two categories. Floriculture is the area of horticulture associated with the production and use of flowers, potted plants, and annual bedding plants. Landscape horticulture includes designing plans for landscapes, installing landscapes as specified in the plans, and maintaining the landscapes. A nursery is a place that specializes in starting plants and growing them until they are ready to be transplanted to landscapes. Garden centers are retail outlets that sell plants grown in nurseries, along with garden supplies.

**4. Chinese-English Interpreting**

昆明国际花卉拍卖交易中心有限公司（KIFA），成立于 2001 年，是国家发改委和云南省政府批准设立的唯一认可的花卉拍卖交易中心，它不仅是花卉拍卖的枢纽，更是一个综合性的产业服务平台，涵盖花卉标准制定、新品种研发推广、市场信息服务、技术服务、金融服务以及物流服务。KIFA 肩负着重大的使命，即推动云南“云花”产业链的创新升级，并致力于实现花卉交易的国际化、标准化、信息化和数字化。经过二十载的蓬勃发展，KIFA 已成为亚洲交易规模最大的花卉拍卖市场，全球排名第二。KIFA 现拥有占地 16 万平方米的交易场馆，有两个拍卖大厅、12 座交易大钟和 900 个交易席位，同时在北京、上海、兰州、菏泽等地设有分中心，同步进行拍卖交易。

目前，KIFA 汇聚了 2.5 万个花卉生产者（供货商）会员和 3,000 多个产地批发商（购买商）会员，日均鲜花交易量达 450 至 500 万枝，平均售价达 1.6 元每枝。高峰时，日交易量曾突破 931 万枝，单价更是飙升至 4.8 元每枝。其中，云南玫瑰占据交易量的七成以上，多头玫瑰更是高达九成，满天星也占据八成以上的市场份额。

KIFA 的拍卖交易涵盖了玫瑰、非洲菊、满天星、洋桔梗、康乃馨、绣球等 40 多个品类，超过 3000 个品种的鲜切花，这些花卉通过拍卖交易进入国内各大城市以及泰国、日本、新加坡、俄罗斯、澳大利亚等 40 多个国家和地区。KIFA 的上下游会员带动了近 30 万花农脱贫致富。

在品种引进与保护方面，KIFA 于 2006 年创新推出了“花卉新品种引进及品种保护三方合作”模式以加强保护机制创新。KIFA 联合国内外花卉育种商和种植者，通过签订三方合作协议，解决了国际花卉新优品种引进和专利费收取的难题，使全球最新、最优的花卉品种源源不断地进入中国。

截至 2021 年，已有来自荷兰、法国、德国、以色列等 18 个国家的育种商的 200 多个专利新品种和 22 个国内自主知识产权新品种通过 KIFA 推向中国市场。2021 年，国内外育种商通过 KIFA 的专利品种交易额高达 7 亿元。同时，KIFA 还利用大数据发布品种流行趋势，对市场出现的新品种进行广泛宣传和推广。

### 5. English-Chinese Interpreting

## Speech at the Opening Ceremony of Kunming International Flower Exposition

Distinguished Guests, Ladies and Gentlemen,

It is a great honor to be here to attend this opening ceremony of the Kunming International Flower Exposition. I feel very grateful that the organization made the Netherlands the country of honor.

Yunnan and the Netherlands have a long and strong relationship. Kunming is known as the "City of Eternal Spring". This is the ideal climate for flowers, which is also recognized by the many Dutch growers located around Kunming.

I am proud to say that this year at the Kunming International Flower Exposition, 29 Dutch companies have joined together in the Netherlands Pavilion. I would surely invite you to come and have a look. As you probably all know, the Netherlands is known for its flowers. The Netherlands is home to the largest flower auction in the world: Aalsmeer. The Netherlands accounts for about half of the world exports of flowers.

Then you might ask, what is the secret behind this success? A big part of the answer is innovation, sustainability and last but not least is cooperation. Specifically in the horticulture the Netherlands has a long history of cooperation within the sector. Our auctions are a nice result of this cooperation. All our auctions, whether they deal with flowers, vegetables or fruits, are part of cooperatives owned by the farmers.

Innovation and sustainability are also key concepts in the agricultural policy of the Netherlands. Innovation in products, technology, organization and business has proven to be essential for the competitiveness and long term sustainability of the horticultural sector in the Netherlands. The innovation in the agricultural sector is a result of a close cooperation between research and private sector and the government, and the role of the government is to create the right enabling environment in which researchers and private sector can play their role. That is why horticulture in the Netherlands has an internationally recognized top ranking and is capable of providing clients worldwide with high quality products, services, education and technology.

The Netherlands is world leading in flower breeding, and most of these companies are also present here in China to develop local specific varieties. Developing new varieties demand significant upfront investments. Seed companies spend more than 30% of their turnover in R&D for developing the varieties of the future. Therefore, we welcomed the improved intellectual property protection under the new Chinese seed law.

Last year, the embassy organized together with the Yunnan Flower Technology Training and Promotion Center an online workshop on Plant Variety Protection. Do you know how many people participated? 83,000 people! This shows that the plant variety protection is a joint interest for all breeders.

The ties between our countries reach far beyond the excellent relations between our governments. China and the Netherlands have built an impressive agricultural trade and investment relationship. With more Dutch companies investing in China's agriculture, more Chinese companies are investing in Dutch food agriculture sector than ever before. And this makes perfect sense, because in many aspects we face similar challenges. And we have similar ambitions to make our countries a better place.

I'm looking forward to an interesting and colourful Flower Exposition.

Thank you.

## Related knowledge

### Plant Variety Protection

Plant breeders innovate and invest to bring new varieties to the world and this continues to be ever more important in these uncertain times. With significant variations in plant variety rights systems around the world, challenges remain for companies to decide how best to protect the investments they make, consider the impact of these decisions, and reflect on which jurisdiction might be most affected. Plant Breeders' Rights (PBRs) are used to protect new varieties of plants that are distinguishable, uniform, and stable. A PBR is legally enforceable and gives you, the owner, exclusive rights to commercially use it, sell it, direct the production, sale, and distribution of it, and receive royalties from the sale of plants.

In many countries PBR lasts for up to 25 years for trees and vines and 20 years for other species. To be eligible for protection, the new variety must be shown to be distinct, uniform, and stable. In a comparative trial the variety must be clearly distinguishable from any other variety, the existence of which is a matter of common knowledge.

In Australia, for example, a new variety may be sold for up to 12 months in Australia, and 4 years overseas, yet still remain eligible for PBRs. These times may be different in other countries. The rights do not cover the use of a grower's crop (the grower does not have to pay a royalty on the crop produced) nor do they extend to retention by growers of seed for the production of another crop or the use of the variety in plant breeding.

A PBR gives you exclusive rights to: produce or reproduce the plant material; condition the plant material for the purpose of propagation; offer the plant material for sale; sell the plant material; import and export the plant material; stock the plant material.

The International Convention for the Protection of New Varieties of Plants, or UPOV Convention, was adopted on December 2, 1961, following a diplomatic conference held in Paris, France. It came into force on August 10, 1968, after having been ratified by the United Kingdom, the Netherlands, and Germany. It was subsequently revised in 1972, 1978, and 1991. The United States is a signatory to the 1991 Act of the UPOV Convention.

The UPOV Convention is administered by the International Union for the Protection of New Varieties of Plants (UPOV), an intergovernmental organization with headquarters in Geneva, Switzerland. The mission of UPOV is to provide and promote an effective system of plant variety protection, with the aim of encouraging the development of new varieties of plants, for the benefit of society.

The UPOV member countries hold annual meetings of the UPOV Council, a permanent body under the UPOV Convention. Other UPOV bodies include the Consultative Committee, the Administrative and Legal Committee, and the Technical Committee, made up of several Technical Working Parties (TWPs) covering several agricultural sectors. The TWPs meet periodically to share and discuss technical observations and advancements in agricultural sectors, which help to standardize examination procedures used by the members. As of June 2021, UPOV has 77 members.

## 口译职业道德规范

口译员，作为语言服务的从业者，致力于在不同语言背景的个体或群体间传递口头信息，以促进有效的信息交流。口译工作不仅需要口译员具备专业技能，更需要口译员严格遵循《译员职业道德准则与行为规范》，确保服务的专业性和质量，为客户提供优质的翻译服务。口译职业道德包括热爱祖国、保守机密、保持中立、量力而行和遵守角色界限。

**1. 口译职业道德**

（1）热爱祖国：在国际舞台上，口译员的一举一动、一言一行代表的不仅是个人，更是国家的形象。因此，口译员必须保持高尚的道德品质，洁身自好，不谋私利，不做任何有损国家尊严和人格的事情。

（2）保守机密：口译员在接触敏感信息时，如讲稿、合同、幻灯片乃至内部机密文件等，必须严格保密，未经授权不得向任何第三方透露，包括其亲朋好友。更不应利用口译工作之便借机谋取个人利益。

（3）保持中立：无论服务对象是谁，口译员都应保持客观中立，准确传达信息，不偏袒任何一方。

（4）量力而行：口译员应评估自己的能力和资源，即使薪酬诱人，不轻易接受超出自己能力范围的任务，以确保服务质量和自身声誉。如较为陌生的领域，口译员缺乏足够的背景知识；或者任务比较紧急，口译员几乎没有时间做充分准备；或在同一时间接受一个以上口译任务。以上几种情形均为超出口译员能力范围的情况。口译员要以最佳精神面貌和工作状态提供语言服务，确保能高质量的完成任务，否则会有损自己的声誉，失去客户的信任。

（5）遵守角色界限：口译员的主要职责是搭建沟通桥梁，使得沟通顺利进行。会谈的内容和走向则由会谈各方来把握。口译员不应代替发言人回答问题或参与协商。这种越界行为不符合译员职业道德准则。

**2. 口译礼仪**

口译礼仪包括遵时守约、着装得体、仪容整洁、交谈得体等要求。

（1）遵时守约：外事活动中，时间观念至关重要。口译员应准时或提前到达活动现场，以示尊重。

（2）着装得体：根据活动性质选择合适的服装，服装应朴素大方、整洁美观。男士出席庆典仪式、正式宴会或会见外宾等外事活动时，可着中山装、西装等正式服装。参观、旅游时可穿便服，如穿西装可不系领带。西装袖子的长度以到手腕为宜，要配好衬衫。衬衫的领子要挺括，系扣。衬衫的下摆要塞在裤子里。

女士服装的式样、颜色可多样化，可根据具体情况着合适的服装，应端庄大方，不宜太花枝招展或珠光宝气。按季节与活动的性质不同，可穿西服、民族服装、中式上衣配长裙或长裤、旗袍和连衣裙等。夏季可穿长、短衬衫，配裙子或长裤。一般外事场合，穿皮鞋、布鞋、凉鞋均可，但不应光脚穿凉鞋或裸露肩膀。在国外参加外事活动，还应尊重东道主的要求和当地的习惯。

（3）仪容整洁：保持整洁的外貌，如理发，刮胡，修剪指甲、鼻毛等，内衣、外衣要保持整洁，特别是衣领口要干净。皮鞋要擦亮，布鞋要洗净。总之，要保持外貌整洁、美观。在参加外事活动时，要注意个人卫生习惯。不要当着他人的面擤鼻涕、掏鼻孔、搓泥垢、挖眼屎、打哈欠、修指甲、剔牙齿、挖耳朵等。咳嗽、打喷嚏时，应用手捂住口鼻，面向一旁，避免发出大声。不随地吐痰，不随地丢果皮纸屑。

（4）交谈得体：与人交谈时，表情自然、态度亲切，表达得体。说话时可适当做些手势，避免过于夸张的手势。在参与他人谈话时，应礼貌地打招呼并尊重他人的隐私。不宜与对方离得太远，但也不要靠得过近。若有事需与某人谈话，应待其说完。有人与自己主动说话，应乐于交谈。第三者参与谈话，应以握手、点头或微笑表示欢迎。发现有人欲与自己谈话时，可主动询问。

通过以上规范，口译员可以确保自己在工作中展现出高度的专业素养和道德风范，为客户提供优质的翻译服务。

## 1. Sentence Interpreting

（1）国际植物新品种保护联盟（UPOV）是总部位于瑞士日内瓦的一个政府间组织。

（2）UPOV 是在《植物新品种保护国际公约》的基础上设立的，该公约于 1961 年在巴黎正式通过，并经历了 1972 年、1978 年和 1991 年的多次修订。

（3）UPOV 致力于构建并执行高效的植物品种保护体系，以鼓励新植物品种的培育，从而造福于整个社会。

（4）虽然“物种”一词在植物界中通常用于描述植物分类的基本单元，但在一个物种内部，往往存在多种具有不同特性的培育品种。农民和种植者通常需要选择那些具有特定特征且适应其种植环境和实践的植物品种。

（5）植物品种是指一个更为具体、定义更精确的植物类群，它们是从某一物种中筛选出来的、具有一系列共同特征。

（6）After thirty years of development, Yunnan’s flower industry has taken the lead in terms of industry scale, market supply, variety, and market share in China.

（7）In recent years, the upgrading of flower consumption has promoted household and daily consumption of fresh flowers. The immense consumer potential of the Chinese flower consumption market will continue to drive the rapid development of Yunnan’s flower industry.

（8）With the advent of the 5G era, digital transformation and upgrading of the flower industry are imperative.

（9）Supported by favorable policies such as import and export tax exemptions, as well as strong promotion of major projects such as national industrial clusters and the provincial-level “one county, one industry” demonstration project, Yunnan’s flower industry has embarked on a new path of high quality and efficiency.

（10）The flower industry in Yunnan has achieved success in the introduction, breeding, and promotion of new quality varieties, as well as in soilless cultivation, marketing, and brand building.

## 2. Chinese-English Interpreting

### Speech Delivered at the 21st China Kunming International Flower Exhibition

Distinguished guests, ladies and gentlemen:

Good morning, everyone!

I am very pleased to be present at the 21st China Kunming International Flower Exhibition. On behalf of the China Flower Association, I would like to extend my warm congratulations on the grand opening of the flower exhibition! I would like to express appreciation to leaders at all levels and friends from all circles who have long been caring and supporting the development of China’s flower industry!

Kunming has superior natural conditions and has long been famous for its flower industry, making it the flower capital renowned worldwide. The China Kunming International Flower Exhibition has been successfully held for 20 years and has played an important role in showcasing new achievements, promoting exchanges and cooperation, and facilitating flower trade in China's flower industry.

Yunnan Provincial Party Committee and the Provincial Government have attached great importance to the development of the flower industry. With relentless efforts, the acreage of flower cultivation and sales have long been leading in China, and we have explored many good experiences and practices, playing a leading role in demonstration nationwide.

Dear comrades, China has abundant floral plant resources, a long history of flower cultivation, and a profound cultural heritage of flowers. After years of efforts, China's flower industry has made significant progress and China has become the world's largest producer of flowers and an important consumer of flowers. However, compared with developed countries in the flower industry and the demands of building a beautiful China and the people's pursuit of a better life, there is still a large gap.

In last November and this May, specific requirements and goals are put forward explicitly by the National Forestry and Grassland Administration (NAFGA) and the Ministry of Agriculture and Rural Affairs, including "The goal of promoting the high-quality development of the flower industry" "Focusing on building a modern flower industry system with a reasonable strategic layout, a complete industrial chain, an active innovation iteration, high efficiency and good benefits, and coordinated regional development", and "The goal of building a nation with strong flower industry by 2035". For the flower industry development, we must plan systematically and promote it guided by the three directions of "High-quality development" "Modern flower industry system" and "Building a nation with strong flower industry".

Building a modern flower industry network is a must for achieving high-quality development of the flower industry and building a nation with strong flower industry. It is an underpinning for the modernization of the flower industry and a new mission delegated to the flower industry in the new era. Faced with new situations and requirements for the flower industry, we must enhance the preservation of floral germplasm resources, establish a sound conservation system and mechanism, and secure the floral germplasm resources in China; we must strengthen independent innovation in flower seed industry, establish an innovation system for flower seed industry, and enhance the level of independent development of flower varieties in China; also, we must shoulder the historical mission given to the flower industry in the new era and better serve the building of ecological civilization and beautiful China.

Yunnan Province has unique natural conditions and abundant floral resources, with a solid foundation for industrial development. In the new era of promoting high-quality development of the flower industry. It is our earnest expectation that Yunnan will be a leader and a demonstration in promoting the building of a modern flower industry system and creating even more brilliant achievements.

Finally, I wish the flower exhibition a complete success!

Thank you.

### 3. English-Chinese Interpreting

当我们深入思考植物的奇妙之处，不禁为它们那将工程学与化学制造完美融合的结构和功能所折服，这远超出了人类的想象。这些高效、协调且多功能的有机体，无需运动部件，仅以自然资源为燃料便能蓬勃生长。环境因素无疑是推动植物进化的关键力量，它们在适应环境生态位的无限变化中展现得淋漓尽致。园丁们应当明白，这种适应并非仅为了装饰或满足他们的审美，而是植物生存和繁衍的必然选择。

掌握不同植物群体在形态、功能和生存策略上的差异背后的科学原理，将极大地提升园丁们的栽培技艺，并加深他们对园艺的热爱。植物与其生长环境之间的关系是动态的，它们能够迅速应对水源供应和温度变化等短期压力，同时也需敏锐感知季节性变化，以启动如从营养生长到开花等重要的生理过程。人类虽已对许多这样的过程有所了解，并将其应用于园艺实践中，但关于这些机制及其触发信号的深入研究，仍然是一个充满挑战且令人兴奋的科学领域，它对园艺师和商业种植者而言具有深远的影响。

园艺学涵盖了植物的生产和利用，旨在满足食用、舒适和美化等多重需求。与园艺学紧密相连的科学领域是植物学，它专注于植物和植物生长过程的研究。园艺科学则专门处理园艺植物的栽培，而将科学应用于园艺的过程则被称为园艺技术。园艺产业可细分为果树学、蔬菜学和观赏园艺学三大领域。果树学关注水果和坚果作物的种植、采收、储存、加工和销售；蔬菜学则涉及蔬菜作物的种植、采收、储存、加工和销售；而观赏园艺则专注于满足美化需求，包括花卉园艺和景观园艺两大类别。花卉园艺专注于花卉、盆栽植物和一年生花卉的生产与使用，而景观园艺则涵盖景观的设计、实施与维护。苗圃是专门培育植物幼苗的场所，植物在这里生长直到能被移植到其他景观区域，而苗圃中心则是销售植物幼苗和园艺用品的零售店。

**4. Chinese-English Interpreting**

The Kunming International Flower Auction Trading Center Co., Ltd. (KIFA) was established in 2001. It is the only flower auction trading center approved by the National Development and Reform Commission (NDRC) and Yunnan Provincial Government. It is a comprehensive industrial service platform for flower auctions and it integrates flower industry standardization, the research variety development and promotion of new varieties, market information services, technical services, financial services, and logistics services. It undertakes the task of promoting the innovation and upgrading of Yunnan flower industry chain, and achieving the internationalization, standardization, informatization, and digitalization of flower trading. After 20 years of development, the Kunming International Flower Auction Center (KIFA) has become the largest flower auction market in terms of sale in Asia and ranks second globally. KIFA currently has a local trading venue of 160,000 square meters in Kunming, with two auction halls, 12 trading clocks, and 900 trading seats. It also conducts simultaneous auction trading in Beijing, Shanghai, Lanzhou, Heze, and other branch centers.

Currently, KIFA has a total of 25,000 flower producers (suppliers) and over 3,000 origin wholesalers (buyers). The daily trading volume of fresh flowers is around 4.5-5 million stems per day, with an average price of 1.6 yuan per stem. The peak trading volume has exceeded 9.31 million stems per day, with a highest average price of 4.8 yuan per stem. Among these, Yunnan roses account for over 70%, multi-headed roses account for over 90%, and gypsophila accounts for over 80%.

Currently, more than 40 categories and over 3,000 varieties of fresh cut flowers, including roses, gerbera, gypsophila, lisianthus, carnations, and hydrangeas, are traded through auctions and are sold to major cities in China, as well as more than 40 countries and regions such as Thailand, Japan,

Singapore, Russia, and Australia. The upstream and downstream members of the Kunming International Flower Auction Center have lifted nearly 300,000 flower farmers from poverty to prosperity.

In 2006, the Kunming International Flower Auction Center implemented the “Three-party Cooperation Model for New Flower Varieties Introduction and for Variety Protection” worldwide, as a way to strengthen innovative protection mechanisms. The Kunming International Flower Auction Center, together with domestic and foreign flower breeding companies and growers, has solved the problem of introducing international new and quality flower varieties and of patent royalty through signing three-party cooperation agreements, as a way to continuously bring the latest and best varieties to China.

As of 2021, more than 200 patented new varieties from 18 breeding companies in the Netherlands, France, Germany, Israel, etc., have been introduced to Chinese market, along with 22 domestically developed new varieties with independent intellectual property rights. In 2021, the transaction volume of patented varieties through KIFA reached RMB 700 million yuan for both domestic and foreign breeding companies. Additionally, big data is utilized to release information on popular variety trends and to promote the introduction of new varieties to the market.

## 5. English-Chinese Interpreting

### 昆明国际花卉博览会开幕式致辞

尊敬的嘉宾，尊贵的女士们、先生们！

非常荣幸能参与此次昆明国际花卉博览会的盛大开幕。我深感荣幸，荷兰被选为本次博览会的荣誉国家。

云南与荷兰之间，早已建立起深厚而持久的友谊。昆明，这座被誉为“春城”的地方，其得天独厚的气候条件为花卉的繁盛生长提供了绝佳的环境，这也得到了众多来自荷兰种植者的认可与赞誉。

我倍感自豪地告诉大家，今年的昆明国际花卉博览会上，有 29 家荷兰公司联手打造了独具匠心的荷兰馆。我热诚地邀请各位嘉宾莅临参观，一睹荷兰花卉的非凡魅力。荷兰，这个以花卉闻名遐迩的国度，拥有全球最大的花卉拍卖市场——阿尔斯梅尔，同时荷兰花卉出口量占据全球半壁江山。

那么，荷兰花卉产业成功的秘诀何在？答案在于创新、可持续性，尤其是紧密的合作。荷兰园艺业拥有悠久的合作历史，我们的拍卖市场便是这一合作的典范。无论是花卉、蔬菜还是水果，我们的拍卖市场都是农民合作社的组成部分，由农民共同拥有和管理。

在荷兰，创新和可持续性被视为农业政策的核心。产品、技术、组织结构和商业模式的创新对于保持荷兰园艺产业的竞争力和实现长期可持续性发展具有举足轻重的地位。这种创新是研究机构、私营部门和政府紧密合作的结晶。政府在其中扮演着创造有利环境、促进各方发挥作用的角色。这也正是荷兰园艺业享誉国际，能够为全球客户提供优质产品、服务、教育和技术的原因。

荷兰在花卉育种领域处于世界领先地位，许多公司在中国设有分支机构，致力于开发适应当地需求的特色品种。新品种的培育离不开巨大的前期投入，种子公司将超过 30%的营业额用于研发，以确保未来品种的竞争力。中国近期修订的《中华人民共和国种子法》在知识产权保护方面的改进，我们表示由衷的欢迎和支持。

去年，我们大使馆与云南花卉技术培训与推广中心联合举办了植物品种保护主题在线研讨会，参会人数高达 83,000 人，这充分说明植物品种保护是所有育种者共同关心的话题和利益所在。

我们两国之间的合作不仅局限于政府层面，农业贸易和投资领域的合作更是硕果累累。荷兰公司在中国农业领域的投资日益增多，同时，中国企业在荷兰食品农业领域的投资也日趋活跃。这是因为我们在许多方面面临着共同的挑战，拥有相似的愿景——那就是让我们的国家变得更加美好。

在此，我预祝本次花卉博览会圆满成功。

谢谢大家！

# Unit 8 Green Energy
# 绿 色 能 源

### What is Renewable Energy

Renewable power is booming, as innovation brings down costs and starts to deliver on the promise of a clean energy future. American sources of solar and wind generation are breaking records and being integrated into the national electricity grid without compromising reliability.

This means that renewables are increasingly displacing "dirty" fossil fuels in the power sector, offering the benefit of lower emissions of carbon and other types of pollution. But not all sources of energy marketed as "renewable" are beneficial to the environment. Biomass and large hydroelectric dams create difficult trade-offs when considering the impact on wildlife, climate change, and other issues. Here's what you should know about the different types of renewable energy sources–and how you can use these emerging technologies in your own home.

Renewable energy, often referred to as clean energy, comes from natural sources or processes that are constantly replenished. For example, sunlight and wind keep shining and blowing, even though their availability depends on time and weather.

While renewable energy is often thought of as a new technology, harnessing nature's power has long been used for heating, transportation, lighting, and more. Wind has powered boats to sail the seas and windmills to grind grain. The sun has provided warmth during the day and helped kindle fires to last into the evening. But over the past 500 years or so, humans increasingly turned to cheaper, dirtier energy sources, such as coal and fracked gas.

Now that we have innovative and less-expensive ways to capture and retain wind and solar energy, renewables are becoming a more important power source, accounting for more than 12 percent of the U.S. energy generation. The expansion in renewables is also happening at scales large and small, from giant offshore wind farms to rooftop solar panels on homes, which can sell power back to the grid. Even entire rural communities (in Alaska, Kansas, and Missouri) are relying on renewable energy for heating and lighting.

## Glossary

| | |
|---|---|
| renewable | 可再生的，可再生资源 |
| biomass | 生物质能，生物燃料 |
| hydroelectric dam | 水力发电站，水电大坝，水电站坝 |
| trade-off | 交易，抵换 |
| peat | 泥煤 |

| | |
|---|---|
| woodsmoke | 木材烟雾 |
| hydrocarbon | 碳氢化合物 |
| wildfire | 野火 |
| wind farm | 风场 |
| wind turbine | 风力涡轮机 |
| displacement | 被迫迁移 |
| species loss | 物种灭绝 |
| asthma | 哮喘 |
| energy-efficient | 高能效的 |
| decarbonize | 脱碳 |
| forestry residue | 森林残留物 |
| photovoltaic | 光伏 |
| photovoltaic effect | 光伏效应 |
| solar panel | 太阳能电池板 |
| grid | 电网 |
| electric vehicle (EV) | 电动汽车 |
| fuel cell vehicle | 燃料电池汽车 |
| scrapped vehicle | 报废的车辆 |
| fiber optic | 光纤 |
| oil refinery | 炼油厂 |
| power plant | 发电厂 |
| lithium | 锂 |
| lithium iron phosphate | 磷酸铁锂 |
| lithium-ion | 锂电子 |

cobalt 钴

electrolyte 电解液，电解质

solvent 溶剂

lithium metal 锂金属

### 1. Sentence Interpreting

（1）风，作为大自然赠予的奇妙之力，其吹向何方、何时起舞、何时静息，皆非人力所能操控。然而，在同一地域内，风的律动却呈现出一定的规律性。

（2）大理地区以其丰富的风能资源成为云南风能资源开发利用的佼佼者。通过不懈的探索和实践，大理在风电管控方面取得了显著的进步。在 2012 年，大理电网的风力发电量就高达 13.88 亿千瓦时，绿色能源的开发利用成效日益凸显。

（3）风功率密度表示在单位面积、单位时间内通过的能量大小，受风速和空气密度的影响，是衡量风电场风能资源潜力的综合指标。

（4）绿色电力主要来源于清洁能源，包括太阳能、风能以及生物质能等。目前，我国的绿色电力主要以太阳能光伏发电和风力发电为主。

（5）面对中国汽车的竞争，欧盟委员会在当地时间 9 月 13 日正式宣布将对中国电动汽车发起反补贴调查。

（6）The most important benefit of solar energy is that it is available every day, and we won’t run out of it. According to scientists, the sun will be shining for at least 5 billion years, providing us with solar energy.

（7）Unlike fossil fuels, solar energy has no harmful impact on the environment. So, when you use solar energy, you're contributing to saving the planet earth.

（8）Fortunately, there are not many moving parts in solar energy systems, and subsequently, they don't require a lot of maintenance cost.

（9）Almost all reliable manufacturers offer a 20-25 years' warranty for their solar panels. You only need to keep the surface of the panels relatively clean.

（10）The more electricity you get from solar panels, the less you're dependent on the grid, and as a result, the more your energy bills will drop.

## 2. English-Chinese Interpreting

### Clean Energy Benefits the Environment

To understand the impact of energy on the environment, it's important to consider how energy use has evolved throughout human history. Prior to the Industrial Revolution, humans used renewable energy. They heated their homes and cooked their food by burning wood or other biomass (peat, for example). They also harnessed hydropower to spin waterwheels and they crushed grains by using windmills.

While not all energy sources were clean, woodsmoke, for example, is highly polluting–they were renewable, meaning they were derived from resources that would not be depleted.

In the 1700s, the Industrial Age dawned with the discovery of how to burn coal to smelt iron. Coal, which is mostly made up of carbonized plant matter (also known as a fossil fuel), was the beginning. It was used to power machinery and used in transportation (trains and steamships). Burning coal introduced carbon into the atmosphere, a direct contributor to a warming planet. Unlike wood or water, coal is nonrenewable; when all the coal is mined, there is no more.

Over the next 200-plus years, fossil fuels in the form of oil and gas also took hold, and there was an explosion in the use and exploitation of natural resources. As with coal, fossil fuels like oil and gas consist of carbon and hydrocarbons, which burn easily and produce a great deal of energy. They also produce carbon byproducts, which cause air pollution and contribute to acid rain and smog.

Clean energy is good for the environment because it does not produce the harmful emissions that lead to climate change. Climate change–including extreme weather and sea-level rise–causes drought, wildfires, flooding, poverty and displacement, food insecurity, health risks, and species loss.

Clean energy technologies improve human health by reducing air pollution and water and soil contamination. A clean energy economy will lower the rates of asthma, heart conditions, and other health issues related to climate change across the nation. It will also support our ecosystems–geographic areas made up of plants, animals, and other organisms that interact with each other and the physical environment–and help sustain life on Earth.

Energy-efficient processes, practices, materials, and products also benefit the environment, because energy efficiency is vital to decarbonizing the economy. Conserving energy, water, and materials strengthens our resources and reduces emissions.

While harnessing power from renewable energy resources reduces carbon dioxide emissions significantly, using biomass resources tackles additional environmental issues. Biomass is derived from plant-and algae-based materials, including food waste and forestry residue. Converting biomass into sustainable fuels for modes of transportation delivers the following environmental benefits: controlling erosion, improving water quality and quantity, increasing biodiversity, and storing carbon in soil.

### 3. Chinese-English Interpreting

#### ① 新能源汽车

新能源汽车主要包括三种类型：纯电动汽车、插电式混合动力汽车和燃料电池汽车。公安部发布的最新统计显示，2022 年全国机动车保有量达 4.17 亿辆，其中汽车 3.19 亿辆；机动车驾驶人达 5.02 亿人，其中汽车驾驶人 4.64 亿人。我国新能源汽车保有量达 1310 万辆，同比增长 67.13%，呈高速增长态势。

全国新能源汽车保有量达 1310 万辆，占汽车总量的 4.10%，扣除报废注销量比 2021 年增加 526 万辆，增长 67.13%。其中，纯电动汽车保有量 1,045 万辆，占新能源汽车总量的 79.78%。2022 年全国新注册登记新能源汽车 535 万辆，占新注册登记汽车总量的 23.05%，与上年相比增加 240 万辆，同比增长 81.48%。新注册登记新能源汽车数量从 2018 年的 107 万辆到 2022 年的 535 万辆，呈高速增长态势。

2023 年前三季度，新能源汽车产业保持强劲发展势头，自主品牌新能源乘用车国内市场销售占比达到 80.2%，新能源汽车出口 82.5 万辆，同比增长 1.1 倍。数据还显示，今年上半年，中国汽车出口量超过日本，成为世界第一大汽车出口国。

#### ②“一带一路”倡议与清洁能源发展合作

文明是建立在基础设施之上的。中国提出的“一带一路”倡议包括基础设施等项目。当许多西方国家还固守着陈旧观念时，中国正在把世界重新想象成一个有机网络。

借助“一带一路”倡议，中国与沿线国家共同合作，兴建工业园区、铺设光纤网络、建设电厂与炼油厂，并打造隧道、桥梁、公路、管道和港口等一系列基础设施。中国的投资不仅着眼于确保知识密集型产品与服务的长远发展，更致力于深化国际产能合作。

随着中国的清洁能源技术日趋成熟，“一带一路”倡议正显著推动清洁能源技术的进步。通过构建与亚洲、非洲及欧洲等地的全面贸易网络，中国正积极发展可再生能源技术，涵盖水力发电、海洋工程及电动汽车等多个领域。尽管中国对化石燃料的需求依然庞大，但未来二十年，预计中国将在低碳发电和其他清洁能源技术上投资超过 6 万亿美元。这一举措实际上是中国实施的一项长远战略，旨在重塑全球能源基础设施。

中国在清洁能源基础设施领域的投入已显著超越美国和欧盟的总和。根据国际能源署的数据，中国目前拥有全球三分之一的风力发电能力，世界十大风力涡轮机制造商中有四家来自中国，全球十大太阳能电池板制造商中有六家为中国企业，同时中国还占据全球四分之一的太阳能发电能力。中国正积极通过构建绿色“一带一路”，为全球超过约 40 亿的消费者带来绿色、低碳且可持续的发展前景。

国际能源署称，中国的清洁能源规划与投资已成为当前推动全球低碳能源技术发展的核心动力。据美国能源部估算，中国已投入高达 470 亿美元支持太阳能电池板制造。此外，中国不仅建立了全球最大的碳交易市场，还在核电站建设领域取得世界领先地位，其电动汽车的销量更是超越全球其他地区总和。

从长远视角来看，预计到 2023 年，可再生能源将占据全球发电量增长的 70% 以上。鉴于全球约半数的能源可能源自太阳能和风能，中国展现出了成为世界上首个清洁能源超级大国的巨大潜力。

中国在大型基础设施建设项目上的丰厚经验，对推动清洁能源技术在发展中国家的广泛应用具有显著助力。为此，中国积极发行绿色债券，专项投资于清洁能源、清洁交通、资源保护及循环利用、污染防治以及节能环保等多个领域的基础设施建设。

清洁能源技术无疑是 21 世纪的重要产业之一，而“一带一路”倡议覆盖了全球可再生能源的主要市场。这一共同建设的倡议正在推动围绕清洁能源技术的长期战略转变，有望重塑全球能源格局。

## Related knowledge

### What's Next for Batteries

Every year the world runs more and more on batteries. Electric vehicles passed 10% of global vehicle sales in 2022, and they're on track to reach 30% by the end of this decade. Policies around the world are only going to accelerate this growth: recent climate legislation in the US is pumping billions into battery manufacturing and incentives for EV purchases. The European Union, and several states in the US, passed bans on gas-powered vehicles starting in 2035.

The transition will require lots of batteries–and better and cheaper ones. Most EVs today are powered by lithium-ion batteries, a decades-old technology that's also used in laptops and cell phones. All those years of development have helped push prices down and improve performance, so today's EVs are approaching the price of gas-powered cars and can go for hundreds of miles between charges. Lithium-ion batteries are also finding new applications, including electricity storage on the grid that can help balance out intermittent renewable power sources like wind and solar.

But there is still lots of room for improvement. Academic labs and companies alike are hunting for ways to improve the technology–boosting capacity, speeding charging time, and cutting costs. The goal is even cheaper batteries that will provide cheap storage for the grid and allow EVs to travel far greater distances on a charge.

At the same time, concerns about supplies of key battery materials like cobalt and lithium are pushing a search for alternatives to the standard lithium-ion chemistry.

In the midst of the soaring demand for EVs and renewable power and an explosion in battery development, one thing is certain: batteries will play a key role in the transition to renewable energy. Here's what to expect in 2023.

**A radical rethink**

Some dramatically different approaches to EV batteries could see progress in 2023, though they will likely take longer to make a commercial impact.

One advance to keep an eye on this year is in so-called solid-state batteries. Lithium-ion batteries and related chemistries use a liquid electrolyte that shuttles charge around; solid-state batteries replace this liquid with ceramics or other solid materials.

This swap unlocks possibilities that pack more energy into a smaller space, potentially improving the range of electric vehicles. Solid-state batteries could also move charge around faster, meaning shorter charging times. And because some solvents used in electrolytes can be flammable, proponents of solid-state batteries say they improve safety by cutting fire risk.

Solid-state batteries can use a wide range of chemistries, but a leading candidate for commercialization uses lithium metal. QuantumScape, for one, is focused on that technology and raised hundreds of millions in funding before going public in 2020. The company has a deal with Volkswagen that could put its batteries in cars by 2025.

But completely reinventing batteries has proved difficult, and lithium-metal batteries have seen concerns about degradation over time, as well as manufacturing challenges. QuantumScape announced in late December it had delivered samples to automotive partners for testing, a significant milestone on the road to getting solid-state batteries into cars. Other solid-state-battery players, like Solid Power, are also working to build and test their batteries. But while they could reach major milestones this year as well, their batteries won't make it into vehicles on the road in 2023.

Solid-state batteries aren't the only new technology to watch out for. Sodium-ion batteries also swerve sharply from lithium-ion chemistries common today. These batteries have a design similar to that of lithium-ion batteries, including a liquid electrolyte, but instead of relying on lithium, they use sodium as the main chemical ingredient. Chinese battery giant CATL reportedly plans to begin mass-producing them in 2023.

Sodium-ion batteries may not improve performance, but they could cut costs because they rely on cheaper, more widely available materials than lithium-ion chemistries do. But it's not clear whether these batteries will be able to meet needs for EV range and charging time, which is why several companies going after the technology, like US-based Natron, are targeting less demanding applications to start, like stationary storage or micromobility devices such as e-bikes and scooters.

Today, the market for batteries aimed at stationary grid storage is small–about one-tenth the size of the market for EV batteries, according to Yayoi Sekine, head of energy storage at energy research firm BloombergNEF. But demand for electricity storage is growing as more renewable power is installed, since major renewable power sources like wind and solar are variable, and batteries can help store energy for when it's needed.

Lithium-ion batteries aren't ideal for stationary storage, even though they're commonly used for it today. While batteries for EVs are getting smaller, lighter, and faster, the primary goal for stationary storage is to cut costs. Size and weight don't matter as much for grid storage, which means different chemistries will likely to win out.

One rising star in stationary storage is iron, and two players could see progress in the coming year. Form Energy is developing an iron-air battery that uses a water-based electrolyte and basically stores energy using reversible rusting. The company recently announced a $760 million manufacturing facility in Weirton, West Virginia, scheduled to begin construction in 2023. Another company, ESS, is building a different type of iron battery that employs similar chemistry; it has begun manufacturing at its headquarters in Wilsonville, Oregon.

**Shifts within the standard**

Lithium-ion batteries keep getting better and cheaper, but researchers are tweaking the technology further to eke out greater performance and lower costs.

Some of the motivation comes from the price volatility of battery materials, which could drive companies to change chemistries. "It's a game of cost,"

Sekine says.

Cathodes are typically one of the most expensive parts of a battery, and a type of cathode called NMC (nickel manganese cobalt) is the dominant variety in EV batteries today. But those three elements, in addition to lithium, are expensive, so cutting some or all of them could help decrease costs.

This year could be a breakout year for one alternative: lithium iron phosphate (LFP), a low-cost cathode material sometimes used for lithium-ion batteries.

Recent improvements in LFP chemistry and manufacturing have helped boost the performance of these batteries, and companies are moving to adopt the technology: LFP market share is growing quickly, from about 10% of the global EVs market in 2018 to about 40% in 2022. Tesla is already using LFP batteries in some vehicles, and automakers like Ford and Volkswagen announced that they plan to start offering some EV models with the LFP batteries too.

## 1. Sentence Interpreting

(1) Wind is a magical power bestowed by nature, and its direction, timing, and cessation are beyond human control. However, in the same region, the rule of wind can be observed and followed.

(2) Dali has abundant wind energy resources and is a typical area for developing and utilizing wind energy resources in Yunnan. Due to proactive exploration and practical implementation, the level of wind power management and control in Dali has significantly improved. In 2012 alone, the wind power generation of Dali's power grid reached 1.388 billion kilowatt-hours, demonstrating the effectiveness of green energy development and utilization.

(3) Wind power density refers to the amount of energy passing through a unit area in a unit of time, which contains the influences of wind speed distribution and air density. It is a comprehensive indicator of wind energy resources in a wind field.

(4) Green electricity mainly derives from clean energy, including solar energy, wind energy, biomass energy, etc. Currently, solar photovoltaic power generation and wind power generation are the main sources of green electricity in China.

(5) Faced with competition from cars made in China, the European Commission officially announced on September 13th local time that it would

initiate an anti-subsidy investigation into Chinese electric vehicles.

（6）太阳能的核心优势在于其每日的可持续利用性，且资源无穷无尽。科学家们指出，太阳预计还能为我们持续提供至少长达约 50 亿年之久的能量。

（7）相较于化石燃料，太阳能对环境无任何负面影响。因此，选择使用太阳能，您便是为保护地球环境贡献一份力量。

（8）幸运的是，太阳能系统因移动部件有限，从而减少了维护成本的需求。

（9）几乎所有信誉良好的制造商都能为他们的太阳能电池板提供长达 20 至 25 年的保修期。您只需保持电池板表面的相对清洁，即可享受这一保障。

（10）通过太阳能电池板获取的电力越多，对电网的依赖就越小，从而能源账单也会相应大幅减少。

**2. English-Chinese Interpreting**

**清洁能源助力环境保护**

为了深入理解能源对环境的影响，追溯能源使用在人类历史中的演变至关重要。在工业革命之前，人类已采用可再生能源。他们依赖燃烧木材或其他生物质（如泥炭）进行取暖和烹饪，同时，水力被用于驱动水车，而风车则用于谷物碾磨。

尽管不是所有能源都具备清洁性，比如木材燃烧产生的烟雾就是高污染性的，但这些能源是可再生的，源自不会耗尽的自然资源。

在 18 世纪，工业时代的序幕随着燃煤炼铁技术的发现而缓缓拉开。煤炭，作为这一时代的标志性能源，主要由碳化植物物质（即化石燃料）构成，它不仅驱动了机械运转，还推动了交通运输的革新，如火车和蒸汽船。然而，煤炭的燃烧过程向大气进行了碳排放，直接加剧了地球变暖的现象。与木材或水资源不同，煤炭是一种不可再生的资源，一旦开采殆尽，便无法再生，标志着煤炭资源的枯竭。

在接下来的两百多年里，石油和天然气等化石燃料逐步占据了能源的主导地位，推动了对自然资源的爆炸性开发和利用。这些化石燃料，与煤炭相似，主要由碳和碳氢化合物构成，不仅易燃而且能量密集。然而，它们的燃烧过程同样会产生大量的碳副产品，这些排放物不仅导致空气污染，还可能引发酸雨和雾霾等环境问题。

清洁能源对环境具有显著益处，因为它不产生导致气候变化的有害排放物。气候变化，包括极端天气和海平面上升，可能引发严重的后果，如干旱、野火频发、洪水泛滥、贫困和流离失所加剧、粮食危机、健康风险上升以及物种丧失。

清洁能源技术通过减少空气污染、水和土壤污染，显著改善人类健康。发展清洁能源经济将有效降低全国范围内因气候变化而引发的哮喘、心脏病变等健康问题的发病率。同时，清洁能源还能维持生态系统的健康，这些生态系统由植物、动物及其与物理环境的相互作用构成，对于维持地球上的生命至关重要。

高效能的加工、实践、材料和产品均对环境大有裨益，因为能源效率是实现经济脱碳的关键。通过节约能源、水资源等原材料，我们能够增强资源可持续性并减少排放。

利用可再生能源能够显著减少二氧化碳排放，而生物质资源的利用则能有效解决其他环境问题。生物质资源源于植物、藻类等，包括食余残渣和林业残渣。将生物质转化为可持续燃料应用于交通运输具有多重优势：不仅有助于控制土壤侵蚀，还能改善水质和增加水量，促进生物多样性，并将碳有效地储存在土壤中。

**3. Chinese-English Interpreting**

**① New Energy Vehicles**

New energy vehicles mainly include three types: electric vehicles, plug-in hybrid electric vehicles, and fuel cell vehicles. According to the latest statistics released by the Ministry of Public Security, the total number of motor vehicles in China has reached 417 million by 2022, including 319 million cars; the number of motor vehicle drivers has reached 502 million, including 464 million car drivers. The number of new energy vehicles in China has reached 13.1 million, with a year-on-year growth of 67.13%, showing a rapid growth trend.

The total number of new energy vehicles in China has reached 13.1 million, accounting for 4.10% of the total number of vehicles. After deducting the scrapped and deregistered vehicles, it increased by 5.26 million compared to 2021, with a growth rate of 67.13%. Among them, the number of pure electric vehicles has reached 10.45 million, accounting for 79.78% of the total number of new energy vehicles. In 2022, the number of newly registered new energy vehicles nationwide reached 5.35 million, accounting for 23.05% of the total number of newly registered vehicles, an increase of 2.4 million, with a year-on-year growth rate of 81.48%. The number of newly registered new energy vehicles has shown a high-speed growth trend, increasing from 1.07 million in 2018 to 5.35 million in 2022.

In the first three quarters of 2023, the new energy vehicle industry maintained a strong development momentum, with domestic sales of domestically-branded new energy passenger vehicles accounting for 80.2% of the market. The export of new energy vehicles reached 825,000 units, a year-on-year increase of 1.1 times. Data also shows that in the first half of this year, China's automobile exports exceeded Japan, becoming the world's largest automobile exporting country.

## ② The Belt and Road Initiative and Cooperation in Clean Energy Development

Civilization is built on infrastructure connectivity. The Belt and Road Initiative proposed by China, involves infrastructure connectivity and other project cooperation. While many Western countries are still clinging to outdated ideas, China is reconceptualizing the world as an organic network.

Relying on the Belt and Road Initiative, China cooperates with countries along the Belt and Road Initiative to build industrial parks, fiber-optic networks, power plants, oil refineries, as well as infrastructure such as tunnels, bridges, highways, pipelines, and ports. China's investments will ensure the long-term development of knowledge-based products and services and strengthen international capacity cooperation.

China's clean energy technology is becoming mature, and the Belt and Road Initiative will greatly promote the development of clean energy technology. China is building a comprehensive trade network through trade exchanges with Asia, Africa, Europe, and other regions, and through which develop renewable energy technologies, including hydroelectric power, marine engineering, and electric vehicles. Despite China's significant demand for fossil fuels, it is expected to invest over 6 trillion dollars in low-carbon power generation and other clean energy technologies in the next 20 years. In fact, China is implementing a long-term strategy that could reshape the world's energy infrastructure.

China's expenditure on clean energy infrastructure has exceeded the sum of the United States and the European Union. According to data from the International Energy Agency, China already possesses one-third of the world's wind power generation, four out of the top ten wind turbine manufacturers globally, six out of the top ten solar panel manufacturers globally, and one-fourth of the global solar power generation capacity. China is expected to bring green, low-carbon, and sustainable development to over 4 billion consumers through the development of the green Belt and Road Initiative.

As pointed out by the International Energy Agency, China's clean energy planning and investment are key drivers of the current momentum in low-carbon energy technology development. The U.S. Department of Energy estimates that China has invested $47 billion to support the manufacturing of solar panels. China has established the world's largest carbon trading market, leading the world in nuclear power plant construction, and its sales of electric vehicles now surpass the total in other regions globally.

In the long run, renewable energy is expected to account for over 70% of global electricity generation growth by 2023. Considering that approximately 50% of global energy potentially coming from solar and wind, China seems poised to become the first clean energy super-country in the world.

China's extensive experience in large-scale infrastructure construction projects helps accelerate the promotion of clean energy technology in developing countries. To support this, China issues green bonds for infrastructure investments in areas such as clean energy, clean transportation, resource conservation and recycling, pollution prevention and control, energy conservation, and environmental protection.

It is obvious that clean energy technology is an important industry of the 21st century. The Belt and Road Initiative covers major markets of renewable energy worldwide, and the joint development of the Belt and Road Initiative is promoting a long-term strategic shift to clean energy technology, which may reshape the global energy landscape.

# 附录　《云南省生物多样性保护条例》英文版

《云南省生物多样性保护条例》（以下简称《条例》）是全国第一部关于生物多样性保护主题的地方性法规，其内容涉及生物学、生态学、管理学和法律等多学科。

云南地处全球三个生物多样性热点地区及其过渡结合部，主要生物类群物种种数均占全国总数一半以上，是中国生物多样性最丰富的省份，其生物多样性保护具有地方、国家和全球的重要意义。为保护云南丰富的生物多样性，云南生物多样性研究院（西南林业大学）于 2011 年启动了《条例》的起草研究工作。2018 年 9 月 21 日，云南省第十三届人大常务委员会第五次会议审议通过了《条例》，自 2019 年 1 月 1 日起施行。《条例》从 2011 年开始起草到正式颁布施行历时八年，开创了中国生物多样性保护立法的先河，对保护国家生物多样性战略资源具有十分重要的意义。《条例》是在没有上位法的情况下全国首创。为宣传云南省生物多样性保护的法制成果，本书主编承担了《条例》汉译英工作，供联合国生物多样性缔约方大会第十五次会议（UNCBD COP15）展示及交流。

此英文条例可作为生态法规主题背景知识和英文学习补充材料，中文《条例》可在百度百科网站获取。

# REGULATIONS OF YUNNAN PROVINCE ON THE CONSERVATION OF BIOLOGICAL DIVERSITY

(Adopted at the Fifth Session of the Standing Committee of the Thirteenth People's Congress of Yunnan Province on 21st September 2018)

## Chapter I
## General Provisions

**Article 1** In order to conserve biological diversity, safeguard ecological security, advance the building of ecological civilization and promote the sustainable development of economy and society, and to attain the harmonious coexistence between man and nature, the Regulations are formulated in accordance with *The Environmental Protection Law of the People's Republic of China*, other relevant laws and regulations, and in light of the actual situations of Yunnan Province.

**Article 2** These Regulations shall apply to the protection, utilization and management of biological diversity in the jurisdiction of the province.

For the purpose of these Regulations, "biological diversity" means the ecological complex of living organisms (fauna, flora, and microorganisms) interacting with environment and the totality of various ecological processes thereby, including the three levels of diversity of ecosystems, species and genes.

These Regulations shall apply for the conservation of biological diversity unless provided otherwise by laws and superior regulations.

**Article 3** The principles of giving priority to conservation, sustainable use, public participation, benefit sharing, protectors gaining and liability for tort shall be observed in the conservation of biological diversity.

**Article 4** The people's governments at all levels shall be responsible for the conservation of biological diversity in their respective jurisdictions.

Enterprises, public institutions and other producers and business operators shall adopt green production methods with high efficiency in resource utilization and with least impact on biological diversity, prevent and reduce damage to biological diversity, and shall be held responsible for the damage to biological diversity in accordance with the law.

Citizens shall enhance their conservation awareness for biological diversity, adopt low-carbon, recycling and economical ways for a green life, and conscientiously resist acts impairing biological diversity.

**Article 5** The people's governments at or above the county level shall strengthen the ecology and environmental protection, and pollution prevention and control so as to prevent biological diversity from impairment of ecological destruction and environmental pollution.

The people's governments at or above the county level shall adopt economy and technology policies and measures conducive to resource saving and recycling for green development, and shall encourage the development and sustainable use of the biological resources in environmentally sound ways, so as to harmonize the economic and social development with the conservation of biological diversity.

**Article 6** The people's governments at or above the county level shall support scientific and technological research in and rational development and sustainable use of biological diversity, develop and improve incentive mechanisms for science-based and orderly development of the biological industries, promote the informatization for the conservation of biological diversity, and improve the level of conservation, utilization and management of biological diversity.

**Article 7** The people's governments at all levels shall make efforts in publicity and education on the conservation of biological diversity, encourage and support grass-root autonomous bodies, social organizations, enterprises and public institutions, volunteers and local people to carry out dissemination, popularization and education of laws, regulations, and knowledge and skills for the protection and utilization of biological diversity so as to enhance the public awareness and create a sound environment for the conservation of biological diversity.

The news media shall carry out publicity and dissemination about laws and regulations and knowledge related to the conservation of biological diversity, and keep watch over illegal acts through public opinions.

**Article 8** The competent environmental protection authorities of the people's governments at or above the county level shall exercise comprehensive administration concerning the conservation of biological diversity in their respective jurisdictions.

The competent authorities in charge of forestry, agriculture, water resources, housing and urban-rural development, land resources, public health and others concerned of the people's governments at or above the county level shall, in accordance with relevant laws and regulations, exercise supervision and administration of the conservation of biological diversity.

**Article 9** The people's governments at or above the county level shall commend and award institutions and individuals for the outstanding achievements they have made in the conservation of biological diversity, in accordance with the relevant stipulations of the national and provincial people's governments.

## Chapter II
## Supervision and Administration

**Article 10** The environmental protection authorities of the people's provincial, prefectural (municipal) governments shall, in collaboration with other relevant departments, formulate programmes or plans for the conservation of biological diversity, and submit them to the people's governments at the same level for ratification, promulgation and implementation.

The people's governments at or above the county level and the relevant authorities shall align the programmes or plans they formulate with those of the same nature for the conservation of biological diversity, and shall analyze, predict and assess the potential impacts on the conservation of biological diversity and thereby, propose solutions and measures to prevent or minimize the negative impacts.

**Article 11** The people's governments at or above the county level shall mobilize the relevant institutions to carry out general surveys, thematic inventories, and data cataloging of biological diversity resources on a regular basis, and shall develop and improve mechanisms for the inventory, monitoring, assessment, surveillance and prediction, and others for biological diversity.

**Article 12** The environmental protection authority of the provincial people's government shall organize the compilation of the list of biological species, the red list of threatened species and the list of ecosystems in its jurisdiction and release them to the public.

The list of biological species, the red list of threatened species and the list of ecosystems shall be updated in due course based on the findings from the inventories, monitoring and assessment of biological diversity.

**Article 13** The relevant administrations of the protected natural areas shall incorporate the conservation of biological diversity into their mandates and responsibilities, strengthen the conservation and management of biological diversity and infrastructure development, and shall be subject to the guidance and supervision of the competent authorities of the people's governments at the same level.

The protected natural areas referred to in the preceding paragraph include nature reserves, scenic and historic areas, national parks, forest parks, important wetlands, world natural heritage sites, conservancy areas for drinking water sources and conservation zones for aquatic germplasm resources, and any other areas legally designated for the conservation of biological diversity.

**Article 14** The people's governments at or above the county level shall supervise and inspect the protection, utilization and management of biological diversity, develop and improve joint law enforcement mechanism for biological diversity, and implement comprehensive administrative law enforcement to meet the needs for the conservation and administration of biological diversity.

**Article 15** The people's governments at or above the county level shall develop an indemnification mechanism for damage to ecology and environment and a compensation mechanism for ecological protection, and shall establish a mechanism for the conservation of biological diversity under which a tort feasor is liable for damage and a protector is compensated.

The people's governments at or above the county level may organize the valuation of biological diversity and ecosystem services.

**Article 16** The people's governments at or above the county level shall establish and improve a financing mechanism, and increase input in conserving biological diversity, and shall include the expenses for the conservation and management of biological diversity in their financial budget at the same level so as to support infrastructure and capacity building for the conservation of biological diversity.

Social funds shall be encouraged for, supported and channelled to the conservation of biological diversity.

**Article 17** The people's governments at all levels shall strengthen regional cooperation and develop coordinated working mechanisms for information sharing, early-warning and forecast, emergency responses, and other joint efforts for the conservation of biological diversity.

International cooperation in the conservation of biological diversity shall be supported, *inter alia*, policy communication, science and technology exchange, and establishment of transboundary conservation mechanisms. Cooperation projects and talent cultivation conducive to the conservation of biological diversity shall be encouraged.

## Chapter III

## Protection of Species and Genetic Diversity

**Article 18** The people's governments at or above the county level shall strengthen the protection of wildlife species and their genetic resources in the respective jurisdictions, and shall improve the system and network for the conservation of biological diversity, integrating *in-situ* conservation, *ex-situ* conservation and *in vitro* preservation, and shall exert rescuing efforts to protect the rare and endangered species and species with extremely small populations, and shall implement special protection for the species endemic to Yunnan and species whose distribution in China is found only in Yunnan.

**Article 19** The people's governments at or above the county level shall strengthen *in-situ* conservation of the resources of wildlife species, their natural habitats, wild relatives of cultivated plants and close relatives of livestock and poultry, and shall, within their competence, designate the areas with concentrated natural distribution of rare, endangered and other wildlife species as protected natural areas in accordance with the law.

In areas where conditions for delineating protected natural areas are not available, the people's governments at or above the county level may adopt such means as establishing germplasm resources reserves (sites), small areas (spots) for natural habitat protection and designating closed zones (where fishing, mining, logging and grazing are banned) and closure periods (when fishing, mining, logging and grazing are banned) for protection purposes.

**Article 20** The relevant authorities of the people's governments at or above the county level shall, according to the programmes or plans for the conservation of biological diversity, develop and improve the *ex-situ* conservation networks, including botanical gardens, arboretums, breeding centers, wildlife domestication bases, and *in-vitro* preservation facilities, including germplasm resource banks, animal cell banks, and gene banks for livestock and poultry.

Enterprises and public institutions, including universities, colleges, and research institutes, and social organizations shall be encouraged and supported to carry out research in the conservation of biological diversity and to participate in constructing relevant conservation facilities.

**Article 21** If the authorities for commerce and other market-related administrations of the people's governments at or above the county have detected any abrupt changes in the market supply or demand for biological resources and products, they shall timely notify the competent authorities for forestry, agriculture and others related to biological resources. The competent authorities shall then assess the impact of market fluctuations on wildlife resources and strengthen supervising and regulating efforts in accordance with their administrative authorizations, so as to prevent possible damage to wildlife resources.

**Article 22** The collection of, scientific research in and biotechnology development of genetic resources shall not affect the genetic integrity of the populations of wildlife species. The access and utilization of genetic resources may not impair human health, ecological safety and biological diversity, and may not cause damage to the local social production and livelihood. And compensations shall be made for actual damage in accordance with the law.

**Article 23** Any domestic or overseas organization or individual collecting, procuring, and conducting field studies for wildlife species or having them transported or mailed outside of the country, must abide by relevant laws and regulations. The competent authorities shall develop and improve information-sharing mechanisms and circulate the relevant situations in a timely manner.

**Article 24** Any institution or individual may not introduce alien species into nature reserves arbitrarily. Where species introduction is deemed necessary, relevant approval must be obtained as required by law, and field experiments be conducted in compliance with the relevant technical regulations and standards.

**Article 25** It is prohibited to proliferate, release or abandon invasive alien species.

In the case of finding suspected alien species, any institution or individual shall report timely to the administrations of environmental protection, forestry, agriculture, public health and the related management institutions of protected natural areas. The departments or institutions in receipt of the reports shall organize a timely on-site investigation. Once confirmed, new occurrences of invasive alien species in the respective jurisdictions shall be treated promptly and reported to the local people's government and the competent department at the next higher level, and the situations shall be circulated to the adjacent areas.

In the case that the departments or institutions in receipt of the reports lack the professional capability of identifying and treating the case, they shall report the case timely to the departments capable of addressing it, which shall then treat the case in a timely manner in accordance with the procedures specified in the preceding paragraph.

**Article 26** The competent administrations of the people's governments at or above the county level and the relevant management institutions of the protected natural areas, shall, according to their division of mandates and responsibilities, carry out systematic inventory, monitoring, assessment and surveillance of invasive alien species and origins of wildlife-borne epidemics and diseases, and they shall, within their mandates and responsibilities, establish ecological risk surveillance and emergency response mechanisms and carry out prevention and control of the invasive alien species and wildlife-borne epidemics and diseases.

## Chapter IV
## Protection of EcoSystem diversity

**Article 27** The people's governments at or above the county level shall take effective measures to protect the typical natural ecosystems in their respective jurisdictions, establish relevant protected natural areas in accordance with the law, and develop and improve ecological corridors to enhance ecosystem integrity and connectivity.

**Article 28** The provincial people's government shall, within its competence and pursuant to the procedures, designate the priority conservation areas and the ecological conservation redlines, and release them to the public.

Any adjustment to the priority conservation areas and the ecological conservation redlines shall be intended to strengthen protection and be ratified according to relevant regulations.

**Article 29** Environmental impact assessment shall be carried out in accordance with the law for new construction, renovation, or expansion projects and exploitation of natural resources. Special protection, restoration and compensation plans shall be formulated and incorporated into the environmental impact assessment for projects that may cause potential damage or impairment to important ecosystems, species, habitats and natural surroundings.

Any construction projects and exploitation of natural resources in the priority conservation areas shall be assessed in terms of their impact on biological diversity, and such assessments shall form important integral components of the environmental impact assessments.

**Article 30** Where the typical ecosystems and those with important economic and social values, and those native to the province are degraded or damaged, the people's governments at or above the county level shall give priority to formulating restoration plans, and treatment and restoration shall be implemented.

The restoration plans shall include the contents, methods and time-bound planning for treatment and restoration, and may, if deemed necessary, take protective closure measures within a prescribed scope.

## Chapter V
## Public Participation and Benefit Sharing

**Article 31** The environmental protection authorities of the people's governments at or above the county level and others in charge of the conservation and administration of biological diversity shall, within their competence, publicize the relevant information on the conservation of biological diversity in accordance with the law, and improve the procedures for public participation so as to facilitate the participation in and supervision over the conservation of biological diversity by citizens, legal persons and other organizations.

**Article 32** The people's governments at or above the county level shall guide institutions and individuals to use recycled products, substitute products, and other products conducive to the conservation of biological diversity so as to reduce the reliance on wildlife resources.

**Article 33** The people's governments at or above the county level and their competent authorities in charge of environmental protection, forestry, agriculture, health, culture and others concerned shall strengthen the inventory, collection, cataloging and protection of the traditional knowledge, methods and skills for the conservation of biological diversity.

Application for patents, trademarks and geographical indications in accordance with the law for protecting ethnic traditional knowledge and skills related to the utilization of biological diversity shall be encouraged; application for ecological protection areas of ethnic traditional cultures, intangible cultural heritage projects, and their representative transmitters should also be encouraged, as a means of transmission and promotion of the traditional cultures in the protection and utilization of biological diversity.

**Article 34** The people's governments at or above the county level shall develop and improve a mechanism for the access to and benefit-sharing (ABS) of genetic resources and relevant traditional knowledge, so as to share equitably and fairly the economic benefits arising therefrom. Incentive mechanisms integrating the conservation of biological diversity and poverty alleviation should be explored and instituted as a means to encourage the local governments and people to share the benefits from biological diversity.

## Chapter VI

## Legal liabilities

**Article 35** The people's governments at all levels and authorities for the protection and administration of biological diversity shall, in accordance with the law, impose administrative punishments on those in charge or with direct responsibility who commit any of the misconducts as follows. If the misconduct constitutes a crime, the legal liabilities shall be affixed:

(1) Fail to formulate programmes or plans for the conservation of biological diversity as required, or formulate them with falsifications.

(2) Arbitrarily alter programmes or plans for the conservation of biological diversity without authorization.

(3) Fail to take legal actions after finding misconducts in breach of these Regulations.

(4) Conduct any other acts, *inter alia*, abusing authority, committing irregularities, dereliction of duty and playing favoritism.

**Article 36** The management institutions of the nature reserves or the administrative authorities of forestry, agriculture and health and others of the people's governments at or above the county level shall issue an order to whoever introduces alien species arbitrarily into nature reserves to make necessary corrections. The related species resources and illegal income shall be confiscated, and a fine of not less than CNY 50,000 but not more than CNY 150,000 shall be imposed.

**Article 37** The competent authorities of forestry and agriculture of the people's governments above the county level shall, in accordance with their respective mandates and responsibilities, issue orders to those whoever proliferate, release or discard invasive alien species to make corrections. And a fine of not less than CNY 2,000 but not more than CNY 10,000 shall be imposed accordingly. In the case of actual losses, the offenders shall be held liable for indemnification in accordance with the law. If the case constitutes a crime, the legal liabilities shall be affixed in accordance with the law.

**Article 38** Any other acts in violation of the provisions herein shall be dealt with by the competent authorities in accordance with relevant laws and regulations.

## Chapter VII
## Supplementary Provisions

**Article 39** Terminologies used for the purpose of these Regulations:

(1) "Ecosystem" is a dynamic complex of fauna, flora and micro-organism communities and their non-living environment interacting as a functional ecological unit,

(2) "Benefit-sharing" means the fair and equitable sharing of benefits arising from the utilization of parts of biological diversity, such as genetic resources amongst others, including the monetary and non-monetary benefits, scientific and technological achievements, technological transfers and capacity building,

(3) "the list of biological species" means the total number and list of biological species in a region or of a taxonomic group that are verified and catalogued by specialists in the relevant fields as organized by the authoritative institutions specialized in the scientific research in biological diversity, in compliance with the established biological taxonomic systems and data standards, and that are released publicly by the authorities for the conservation and management of biological diversity,

(4) "the red list of threatened species" indicates the status of species

endangerment and is a list of species, of which each and every species in a region is assessed and placed into a category of endangerment by specialists in the relevant fields as organized by the authoritative institutions specialized in the scientific research in biological diversity, in compliance with the internationally recognized and widely used methods and criteria, and that are released publicly by the departments in charge of the conservation and administration of biological diversity,

(5) "the list of ecosystems" means the list of ecosystems in a region that is verified and catalogued by specialists in the relevant fields as organized by the authoritative institutions specialized in the scientific research in biological diversity, in compliance with internationally recognized and widely used methods and criteria, and that is released to the public by the departments in charge of the conservation and administration of biological diversity,

(6) "*In vitro* preservation" means the long-term storage of a part of an organism, including seeds, germs or embryos, tissues, cells, deoxyribonucleic acid (DNA), etc., by using modern technology, in particular, the cryogenic and ultra-cryogenic freezing technology, as a means to preserve the germplasms and genetic resources of biological species,

(7) "Extremely small population species" means a species of wild fauna or flora that, due to its narrow or disjunct distributions and long-term disturbance from external factors, has demonstrated continual population degradation and decrease to extremely small populations and individuals below the stable limits of minimal viable populations (MVPs), so that they may become extinct at short time spans,

(8) "Habitat" means the place or type of site where a living organism or

collection of living organisms is naturally distributed,

(9) “Genetic resources” means any material (excluding human genetic resources) of actual or potential value for humanity from plants, animals, microorganisms or other origins containing functional units of heredity,

(10) “Alien species” means a taxon of species, subspecies or a taxon below the taxonomic unit which has had no natural distribution in the past or present protected natural areas, including all organic parts thereof, gametes or propagules which may survive and reproduce,

(11) “Invasive alien species” means any non-native species that has developed viability in local natural or semi-natural ecosystems and may cause or have caused significant damage to, or adverse impacts on the ecological environment, production or livelihood,

(12) “Priority conservation areas for biological diversity” means the key and critical areas designated for the conservation of biological diversity due to the abundance of species present, the extent of species rarity and endangerment and the representativeness of ecosystem types, and the irreplaceability in a region.

**Article 40** These Regulations shall enter into force on 1st January 2019.

# References
# 参 考 文 献


[1] 李长栓．英语口译案例讲评[M]．北京：外语教学与研究出版社，2013．

[2] 雷天放，陈菁．口译教程[M]．上海：上海外语教育出版社，2013．

[3] 杨柳燕，苏伟．口译教程[M]．2 版．上海：上海外语教育出版社，2016．

[4] 赵军峰．商务英语口译教程[M]．北京：高等教育出版社，2003．

[5] 艾群，陈潇，江雯．新编英语翻译教程[M]．哈尔滨：哈尔滨出版社，2022．

[6] 刘和平．职业口译教学与研究[M]．北京：外语教学与研究出版社，2017．

[7] 梅德明．高级口译教程[M]．上海：上海外语教育出版社，2011．

[8] 仲伟合，何刚强．汉英口译[M]．北京：外语教学与研究出版社，2012．

[9] Washborne, Kelley. Note-taking for Consecutive Interpreting: a Short Course[M]．上海：上海外语教育出版社，2009．

[10] 杨宇明，王娟，王建皓，等．云南生物多样性及其保护研究[M]．北京：科学出版社，2008．